STRUCTURE AND BONDING

Volume 33

Editors:
J. D. Dunitz, Zürich · P. Hemmerich, Konstanz
J. A. Ibers, Evanston · C. K. Jørgensen, Genève · J. B. Neilands, Berkeley · D. Reinen, Marburg · R. J. P. Williams, Oxford

With 85 Figures and 47 Tables

Springer-Verlag
Berlin Heidelberg GmbH 1977

ISBN 978-3-662-15482-3 ISBN 978-3-540-37337-7 (eBook)
DOI 10.1007/978-3-540-37337-7

Library of Congress Catalog Card Number 67-11280

Originally published by Springer-Verlag Berlin Heidelberg New York in 1977
Softcover reprint of the hardcover 1st edition 1977

Contents

STRUCTURE AND BONDING is issued at irregular intervals, according to the material received. With the acceptance for publication of a manuscript, copyright of all countries is vested exclusively in the publisher. Only papers not previously published elsewhere should be submitted. Likewise, the author guarantees against subsequent publication elsewhere. The text should be as clear and concise as possible, the manuscript written on one side of the paper only. Illustrations should be limited to those actually necessary.

SPRINGER-VERLAG

D-6900 Heidelberg 1
P. O. Box 105280
Telephone (06221) 487·1
Telex 04-61723

D-1000 Berlin 33
Heidelberger Platz 3
Telephone (030) 822001
Telex 01-83319

SPRINGER-VERLAG
NEW YORK INC.

175, Fifth Avenue
New York, N.Y. 10010
Telephone 673-2660

Crystal Field Effects in Rare-Earth Intermetallic Compounds

W.E. Wallace, S.G. Sankar and V.U.S. Rao

Department of Chemistry, University of Pittsburgh, Pittsburgh, PA. 15260 (USA)

Table of Contents

I. Scope of the Review

The crystalline electric field interaction plays a very significant role in modifying many properties of metallic rare earth systems – magnetic susceptibility and saturation magnetization, thermodynamic properties such as heat capacity, compressibility and expansivity, transport properties such as electrical and thermal conductivity and superconductivity. Several review articles have recently appeared covering certain specialized aspects of the crystal field interaction (which will be referred to hereafter as CFI). *Cooper* and *Rhyne* (*1*) have discussed various features of the elemental rare earths which are influenced by the CFI. *Malik et al.* (*2*) have summarized observations on Ce and Sm compounds with emphasis on NMR characteristics. These and other aspects of Sm compounds have been treated (*3*) by *de Wijn, Van Diepen* and *Buschow. Lüthi* (*4*) has discussed the effects of the CFI with emphasis on elastic constants, particularly for rare earth pnictides. *Wallace* (*5*) has summarized results obtained pertaining to the CFI for selected rare earth intermetallics through susceptibility and thermodynamic (heat capacity) studies. The purpose of the present review is to update the earlier summary, which covered the field only through 1973 and moreover was in the form of a conference presentation and hence enjoyed only limited circulation.

In Section II we present a brief account of the methods that have been developed to deal with the CFI and exchange interactions in metallic rare earth systems. Usually both of these interactions are involved in varying degrees of relative strength. In Section III a summary account is given of the calculational procedures involved in determining the eigenvalues and eigenfunctions and in relating the observable properties of the system – susceptibility, heat capacity, etc. – to the details of the CFI.

In Section IV we have discussed results for a selection of intermetallic compounds such as RAl_2, RAl_3, RNi_2, RNi_3, RNi_5, RCo_5 and RFe_2. (R here represents a rare earth.) There are certain significant omissions in our coverage. Since Gd^{3+} is in an S state, the CFI interaction for it is insignificant. Hence little attention is paid to Gd compounds in this review. Ce and Yb in intermetallic compounds exhibit variable valency; Ce is found in both the Ce^{4+} and Ce^{3+} states and Yb exists as Yb^{2+} and Yb^{3+}. Because of these complications we have not considered Ce and Yb compounds except in a few instances. Compounds of Sm represent an interesting special case – because of J-mixing effects *(vide infra)* and additionally because of the widespread interest in $SmCo_5$ as an extraordinary permanent magnet material. These compounds are described in a separate section.

The examples cited in Section IV are frequently drawn from work carried out in this laboratory. They have been chosen to illuminate the principles involved. The present review is not indended to be exhaustive, but rather illustrative of principles. Because of space limitations many important topics are not included – behaviour of the elemental rare earths (*1*), amorphous materials (*6*), liquid rare earth systems (*7*), elastic and magnetoelastic properties (*4*), thermal conductivity (*4*), etc. The reader who is interested in these topics is referred to the reviews cited in ref. (*1*), (*4*), (*6*) and (*7*).

II. Introduction

In the rare earth metals and intermetallic compounds, the 4f electrons are well embedded within the core orbitals and therefore do not play a significant role in chemical bonding. The metallic rare earth systems, with a few exceptions, may be treated as an assemblage of tripositive ions located in a sea of conduction electrons. The 4f electrons possess large angular momentum and are therefore responsible for the interesting and variety of magnetic properties exhibited by these rare earth materials. With the progressive filling of the 4f electronic level as one passes from Ce to Lu, the electronic interaction within the atom gives rise to many states with varying and interesting characteristics.

Two other elements are normally considered along with rare earths because of their similar chemical properties: La, which is just before Ce in the sequence, and Y, which comes in the same position as La in the preceding period. These two elements are nonmagnetic and therefore they and their intermetallic compounds serve as "blanks".

A. The Exchange Interaction

In rare earth metals and intermetallic compounds, a strong magnetic coupling, as indicated by their rather high ordering temperatures, is often observed. This is in contrast with the behavior of their oxides and salts. The exchange interaction responsible for this coupling between 4f orbitals is an indirect one involving polarization of the conduction electrons. Such an interaction is oscillatory and has long range character. The theoretical framework on the basis of these ideas was developed by *Ruderman* and *Kittel* (*8*), *Kasuya* (*9*) and *Yosida* (*10*) and is usually known as the RKKY theory. This theory explains a variety of magnetic structures exhibited by rare earth intermetallics. *de Gennes* (*11*), *Kasuya* (*12*), *Elliott* (*13*), *Rocher* (*14*) and *Kittel* (*15*) have dealt with this interaction mechanism in great detail.

B. The Crystal Field Interaction

An array of charges in a crystal produces an electric field at any one ion, the so-called crystalline electric field. The presence of this field causes a Stark splitting of the free ion energy levels which results in a substantial modification of magnetic, electrical and thermal properties of the material. The theory of the crystal field and its interpretation in terms of group theory are originally due to *Bethe* (*16*).

If the origin of the coordinate system is taken at the nucleus of the rare earth ion, an expression for the electrostatic potential at a point (r, θ, φ) near the origin due to the surrounding k ions may be written as

$$V(r, \theta, \varphi) = \sum_k q_k/(\vec{R}_k - \vec{r}) \tag{1}$$

If the rare earth ion has charge q_i at r_i, θ_i and φ_i, its electrostatic energy due to the perturbing potential, V, is

$$\mathscr{H}_{CF} = \sum_i q_i V = \sum_i \sum_k q_i q_k/(\vec{R}_k - \vec{r}_i) \tag{2}$$

This perturbing potential partially lifts the (2 J + 1) degeneracy of the ground multiplet of the free rare earth ion. The resulting eigenvalues and eigenfunctions of the rare earth ions may be evaluated by the straightforward but tedious methods of perturbation theory (*17*). This is seldom done because of the convenience of a calculational method introduced by *Stevens* about 25 years ago *(vide infra)*.

Table 1. Stevens' multiplicative factors associated with equivalent operators for the ground states of rare earth ions and the calculated Hartree-Fock radial integrals $\langle r^n \rangle$ in atomic units of length a_0^{-n} (*20*).

Ion	Ground state	α_J	β_J	γ_J	$\langle r^2 \rangle$	$\langle r^4 \rangle$	$\langle r^6 \rangle$
Ce^{3+}	$^2F_{5/2}$	$\frac{-2}{5 \cdot 7}$	$\frac{2}{3^2 \cdot 5 \cdot 7}$	0	1.200	3.455	21.226
Pr^{3+}	$^3H_4$	$\frac{-2^2 \cdot 13}{3^2 \cdot 5^2 \cdot 11}$	$\frac{-2^2}{3^2 \cdot 5 \cdot 11^2}$	$\frac{2^4 \cdot 17}{3^4 \cdot 5 \cdot 7 \cdot 11^2 \cdot 13}$	1.086	2.822	15.726
Nd^{3+}	$^4I_{9/2}$	$\frac{-7}{3^2 \cdot 11^2}$	$\frac{-2^3 \cdot 17}{3^3 \cdot 11^3 \cdot 13}$	$\frac{-5 \cdot 17 \cdot 19}{3^3 \cdot 7 \cdot 11^3 \cdot 13^2}$	1.001	2.401	12.396
Sm^{3+}	$^6H_{5/2}$	$\frac{13}{3^2 \cdot 5 \cdot 7}$	$\frac{2 \cdot 13}{3^3 \cdot 5 \cdot 7 \cdot 11}$	0	0.883	1.897	8.775
Tb^{3+}	$^7F_6$	$\frac{-1}{3^2 \cdot 11}$	$\frac{2}{3^3 \cdot 5 \cdot 11^2}$	$\frac{-1}{3^4 \cdot 7 \cdot 11^2 \cdot 13}$	0.758	1.44	5.8
Dy^{3+}	$^6H_{15/2}$	$\frac{-2}{3^2 \cdot 5 \cdot 7}$	$\frac{-2^3}{3^3 \cdot 5 \cdot 7 \cdot 11 \cdot 13}$	$\frac{2^2}{3^3 \cdot 7 \cdot 11^2 \cdot 13^2}$	0.726	1.322	5.102
Ho^{3+}	$^5I_8$	$\frac{-1}{2 \cdot 3^2 \cdot 5^2}$	$\frac{-1}{2 \cdot 3 \cdot 5 \cdot 7 \cdot 11 \cdot 13}$	$\frac{-5}{3^3 \cdot 7 \cdot 11^2 \cdot 13^2}$	0.695	1.22	4.5
Er^{3+}	$^4I_{15/2}$	$\frac{2^2}{3^2 \cdot 5^2 \cdot 7}$	$\frac{2}{3^2 \cdot 5 \cdot 7 \cdot 11 \cdot 13}$	$\frac{2^3}{3^3 \cdot 7 \cdot 11^2 \cdot 13^2}$	0.666	1.126	3.978
Tm^{3+}	$^3H_6$	$\frac{1}{3^2 \cdot 11}$	$\frac{2^3}{3^4 \cdot 5 \cdot 11^2}$	$\frac{-5}{3^4 \cdot 7 \cdot 11^2 \cdot 13}$	0.640	1.03	3.45
Yb^{3+}	$^2F_{7/2}$	$\frac{2}{3^2 \cdot 7}$	$\frac{-2}{3 \cdot 5 \cdot 7 \cdot 11}$	$\frac{2^2}{3^3 \cdot 7 \cdot 11 \cdot 13}$	0.613	0.960	3.104

Because of the fact that the 4f electrons are well shielded from the above potential by the outer electrons, the splitting of the energy levels in rare earth ions is relatively small (with a few exceptions such as Sm and Eu ions) compared to spin-orbit interaction. Thus it is often sufficient to perform the calculations employing only the ground multiplet. Under these conditions the quantum mechanics of the interaction are very much simplified. Stevens introduced what has come to be known as the method of operator equivalents (*18*) to deal with the case when only the ground state multiplet is involved. In this method an operator consisting of standard angular momentum operators acting on the angular part of the wave functions of the rare earth ion, which is described in the $|LSJM\rangle$ representation, is determined. Thus we may write for the crystal field term:

$$\mathscr{H}_{CF} = \sum_{n=0}^{n'} \sum_{m=-n}^{n} B_n^m \, O_n^m(\vec{J}) \tag{3}$$

where n' can take a maximum value of n. B_n^m are the crystal field intensity parameters and $O_n^m(\vec{J})$ represent polynomials of the angular momentum operators J_z, J^2, J_+, J_-. The polynomials and the matrix elements associated with several operators for various values of J have been conveniently tabulated by *Hutchings* (*19*). It is to be noted that:

a) for f electrons n' is 6 and for d electrons it is 4;
b) if there is an inversion center there are no odd n terms; and
c) the site symmetry of the rare earth ions reduces the number of terms which have to be taken into consideration.

For ions located at points of very high symmetry the coefficients with the same n may be related to each other.

The B_n^m coefficients are determined in part by the surrounding ions and in part by the radial extension and the total angular momentum of the rare earth ion. B_n^m is evaluated from the expression

$$B_n^m = \langle J \| \theta_n \| J \rangle \langle r^n \rangle (1 - \sigma_n) K_n^m A_n^m \tag{4}$$

where $\langle J \| \theta_n \| J \rangle$ is a reduced matrix element which is usually designated as α_J, β_J and γ_J for $n = 2$, 4 or 6, respectively; $\langle r^n \rangle$ is the expectation value of the nth power of the 4f electron radius. These values are listed in Table 1 for the tripositive rare earth ions (*20*). $(1 - \sigma_n)$ is a factor which takes into account the shielding of the 4f wave functions from the environment by the outer, filled 5s and 5p shells (*21*) Values of the constants K_n^m occurring in Eq. (4) are given in Table 2. A_n^m is a measure of the lattice sum and is controlled by the electrostatic potential generated by the surrounding ions. It is given by the expression

$$A_n^m = (-1)^{m+1} [4\pi/(2n+1)] e^2 \sum_k (Z_k/R_k^{n+1}) Y_n^{-m}(\theta_k, \varphi_k) \tag{5}$$

Table 2. Values of constants K_n^m used in Eq. (4) [a]

$K_2^0 = 1/4(5/\pi)^{1/2}$	$K_2^1 = (15/2\,\pi)^{1/2}$	$K_2^2 = 1/2(15/2\,\pi)^{1/2}$
$K_4^0 = 3/16(\pi)^{1/2}$	$K_4^1 = 3/4(5/\pi)^{1/2}$	$K_4^2 = 3/4(5/2\,\pi)^{1/2}$
	$K_4^3 = 3/4(35/\pi)^{1/2}$	$K_4^4 = 3/8(35/2\,\pi)^{1/2}$
$K_6^0 = 1/32(13/\pi)^{1/2}$	$K_6^1 = 1/8(13\cdot 21/2\,\pi)^{1/2}$	$K_6^2 = 1/32(13\cdot 105/\pi)^{1/2}$
	$K_6^3 = 1/16(13\cdot 105/\pi)^{1/2}$	$K_6^4 = 3/32(13\cdot 14/\pi)^{1/2}$
	$K_6^5 = 3/16(13\cdot 77/\pi)^{1/2}$	$K_6^6 = 1/32(13\cdot 21\cdot 11/\pi)^{1/2}$

[a]) Ref. (*107*).

where the summation is over all the surrounding ions. Z_k is the charge on the kth ion which is located at $(\vec{R}_k, \theta_k, \varphi_k)$.

The eigenvalues and eigenfunctions obtained by diagonalizing the Hamiltonian in Eq. (3), using crystal field intensity parameters obtained from Eq. (4) and Eq. (5), gives a representation of the crystal field interaction according to the point charge model. Although the point charge model is a convenient conceptual framework for assessing the interaction, there are additional effects which may modify the results significantly. The role played by 5 d and 6s conduction electrons in antishielding has been investigated by *Das* and *Ray* (*22*), while the influence of covalent bonding has been discussed by *Jørgensen et al.* (*23*). Further, there is an uncertainty involved in determining the number of neighbors one has to consider in applying the point charge model to the metallic state, and in addition the screening effect of the conduction electrons is difficult to assess. In intermetallic compounds there is often uncertainty as to the charges associated with the non-rare earth partners. It has therefore become customary to treat $A_n^m \langle r^n \rangle$, satisfying the symmetry requirements, as adjustable parameters to be determined from experiment.

III. Calculational Procedures

A. *Methodology for Treating the Interactions*

As noted earlier, the effect of crystal field interaction in rare earth intermetallics results in the modification of magnetic, thermal, electrical and magnetocrystalline anisotropic properties. These properties may be calculated if the energy eigenvalues and eigenfunctions are known for the ions under the influence of the appropriate crystal field. In this section, we briefly discuss methods of calculating the eigenvalues and eigenfunctions and from them the physical properties mentioned above.

1. The Case of a Single J State

Within a manifold of angular momentum J composed of f electron wavefunctions the most general operator equivalent potential with cubic point symmetry may be expressed as

$$\mathscr{H}_{CF}^{\langle 100\rangle} = B_4(O_4^0 + 5\,O_4^4) + B_6(O_6^0 - 21\,O_6^4)\,. \qquad (6)$$

Here the reference axis is taken along the fourfold axis.

Lea, *Leask* and *Wolf* (*24*) in their classical paper provided a detailed treatment of the ions for J values ranging from 2 to 8. For computational convenience the above expression may be rewritten as

$$\mathscr{H}_{CF}^{\langle 100\rangle} = W\left[x\left(\frac{O_4}{F_4}\right) + (1 - |x|)\left(\frac{O_6}{F_6}\right)\right] \qquad (7)$$

where

$$O_4 = O_4^0 + 5\,O_4^4\,,$$
$$O_6 = O_6^0 - 21\,O_6^4\,,$$
$$B_4 = Wx/F_4$$

and

$$B_6 = W(1 - |x|)/F_6\,.$$

x here represents the relative importance of the fourth and sixth order crystal field; by varying x from -1 through 0 to $+1$, B_4/B_6 can be scanned in the range $-\infty$ to $+\infty$. W is a scale factor relating the eigenvalues to the crystal field energy levels. F_4 and F_6 are factors common to all matrix elements (*19*). For a given ion diagonalization of the $(2J + 1) \times (2J + 1)$ matrix provides the eigen energy values and eigenfunctions for a given ratio of B_4 to B_6. As an illustration, the energy levels of Er^{3+} ion under the influence of cubic crystalline field as a function of x are shown in Fig. 1.

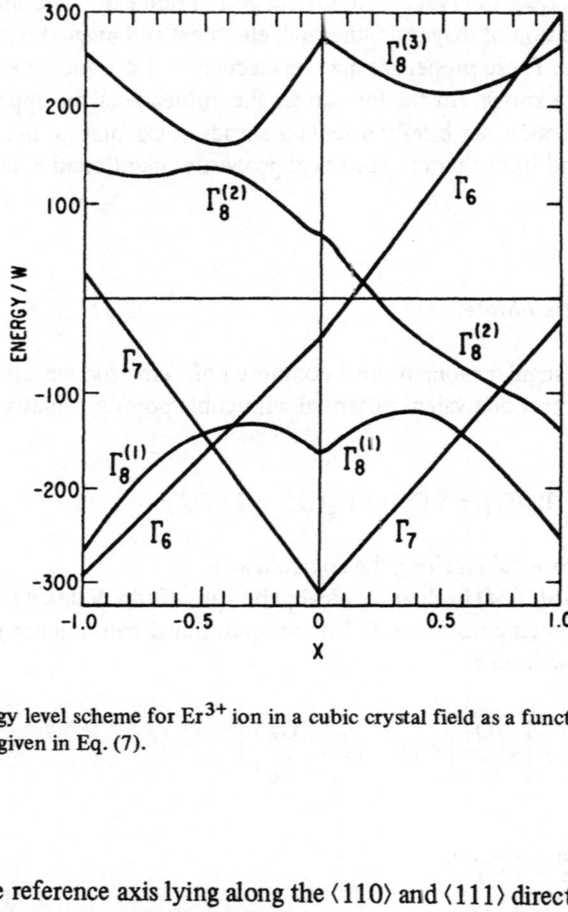

Fig. 1. Energy level scheme for Er^{3+} ion in a cubic crystal field as a function of the crystal field parameter x given in Eq. (7).

For the reference axis lying along the ⟨110⟩ and ⟨111⟩ directions, the crystal field Hamiltonian may be expressed as (*19*)

$$\mathscr{H}_{CF}^{\langle 110\rangle} = -0.25\,B_4(O_4^0 - 20\,O_4^2 - 15\,O_4^4) - \frac{13}{8}B_6\left(O_6^0 + \frac{105}{26}O_6^2 - \frac{105}{13}O_6^4 + \frac{231}{26}O_6^6\right) \tag{8}$$

and

$$\mathscr{H}_{CF}^{\langle 111\rangle} = -\frac{2}{3}B_4(O_4^0 - 20\sqrt{2}\,O_4^3) + \frac{16}{9}B_6\left(O_6^0 + \frac{35\sqrt{2}}{4}O_6^3 + \frac{77}{8}O_6^6\right) \tag{9}$$

Segal and *Wallace* (*25*) reported detailed calculations involving the effect of hexagonal crystal field in rare earth ions. Their Hamiltonian is of the form

$$\mathscr{H}_{CF} = B_2^0 O_2^0 + B_4^0 O_4^0 + B_6^0 O_6^0 + B_6^6 O_6^6 \qquad (10)$$

As in the case of cubic system, for computational convenience, the above Hamiltonian may be rewritten as

$$\mathscr{H}_{CF} = W[(1-|y|)\frac{O_2^0}{F_2} + y\{x\frac{O_4^0}{F_4} + (1-|x|)(O_6^0 + \frac{77}{8}O_6^6)\} \qquad (11)$$

The quantity $\frac{77}{8}$ holds only to the ideal hcp structure (c/a = 1.63).

2. *Effects of J Mixing*

In some ions, such as Sm^{3+} and Eu^{3+}, because of the close proximity of the excited multiplets to the ground state, it is necessary to include one or more of the higher multiplets in evaluating the magnetic and related properties. Recent papers by *Malik et al.* (*2*) and *de Wijn et al.* (*3*) provide excellent information on the details of the calculations. However, we give below some salient features of the effects of J mixing.

In calculations involving higher J multiplets, matrix elements of the crystal field Hamiltonian between states belonging to different J multiplets are needed. Although these can be calculated by the method of operator equivalents extended to elements non-diagonal in J, it is convenient to use a more general approach, utilizing Racah's tensor operator technique (*26*). In this method the crystal field interaction may be written as

$$\mathscr{H}_{CF} = \Sigma N_n^m A_n^m \langle r^n \rangle U_m^n \qquad (12)$$

where N_n^m are normalization factors tabulated by *Weber* and *Bierig* (*27*) and

$$U_m^n = \sqrt{4\pi/(2n+1)}\, Y_n^m \qquad (13)$$

The expression for the matrix elements of U_m^n can be written in terms of the Wigner 3–j and 6–j symbols as follows:

$$\langle \alpha SLJ J_z | U_m^n | \alpha SLJ' J_z' \rangle = (-1)^{J z + S + L} [(2J+1)(2J'+1)]^{1/2} \qquad (14)$$

$$x \begin{pmatrix} J & n & J' \\ -J_z & m & J_z' \end{pmatrix} \begin{Bmatrix} L & J & S \\ J' & L & n \end{Bmatrix} \langle \alpha SL \| U^n \| \alpha SL \rangle$$

where α is used to describe the quantum states completely. The large parenthesis represents the 3–j symbol, the large curly parenthesis represents the 6–j symbol. These have been tabulated by *Rotenberg et al.* (*28*). The last term in Eq. (14) represents reduced matrix elements which have been conveniently tabulated by *Nielson* and *Koster* (*29*). Thus by using the reduced matrix elements and the 3j, 6j symbols, the required matrix elements of the crystal field Hamiltonian [Eq. (12)] may be evaluated. Diagonalization of the matrix thus constructed provides the eigenvalues and eigenfunctions needed in calculating the magnetic and related properties as described later.

At this point, it is pertinent to point out that in recent years the influence of the higher multiplets on the crystal field splittings in ions such as Nd^{3+} and Dy^{3+} has been investigated. *Schlesinger* and *Nara* (*30*) examined the effect of inclusions of three J multiplets in Dy^{3+} in cubic environment. *Furrer* and *Tellenbach* (*31*) calculated the magnetic moment in some neodymium compounds and concluded that results are in much better agreement with experiment than if the higher multiplets were ignored. In general, it is noted that for systems with a large sixth order contribution to the crystal field potential, the effect of higher multiplets is significant.

3. *Effect of External or Exchange Fields*

As noted above the rare earth intermetallics exhibit reasonably strong crystal fields. The influence of external fields and exchange fields vary to a great extent – sometimes their effects are negligible while in other cases they have dramatic effects. Examples of this behavior will be given in a later part of this paper. The effect of exchange field may be expressed as follows:

$$\mathscr{H}_{\text{mag}} = 2\,\mu_B\,\vec{S}\cdot\vec{H}_{\text{exch}} \tag{15}$$

where μ_B is the Bohr magneton, $\vec{S}$ the spin of the rare earth ion and $\vec{H}_{\text{exch}}$ the exchange field acting on the spin of the rare earth ion. When the effect of higher multiplets is ignored in the crystal field Hamiltonian, the above expression may be simplified as

$$\mathscr{H}_{\text{mag}} = 2\,(g_J - 1)\,\mu_B\,\vec{J}\cdot\vec{H}_{\text{exch}} \tag{16}$$

In systems with uniaxial symmetry the field may be applied along any one of the two possible directions, namely, along the *c*-axis and along a direction lying in the basal plane. Under these conditions the above equation may be expressed as

$$\mathscr{H}^{\parallel}_{\text{mag}} = 2\,(g_J - 1)\,\mu_B\,J_z\,H_{\text{exch}} \tag{17}$$

and

$$\mathscr{H}^{\perp}_{\text{mag}} = 2\,(g_J - 1)\,\mu_B\,J_x\,H_{\text{exch}} \tag{18}$$

The total Hamiltonian of a system thus consists of a crystal field term and an exchange (with or without external) field and may be expressed as

$$\mathscr{H}^{\parallel} = \mathscr{H}_{\text{CF}} + \mathscr{H}^{\parallel}_{\text{mag}} \tag{17a}$$

$$\mathscr{H}^{\perp} = \mathscr{H}_{\text{CF}} + \mathscr{H}^{\perp}_{\text{mag}} \tag{18a}$$

In cases where the higher multiplets have to be included, it is appropriate to write the above expressions as

$$\mathscr{H}^{\parallel}_{\text{mag}} = 2\,\mu_B\,S_z\,H_{\text{exch}} \tag{19}$$

$$\mathscr{H}^{\perp}_{\text{mag}} = 2\,\mu_B\,S_x\,H_{\text{exch}} \tag{20}$$

The relevant matrix elements for the above cases are (32)

$$\langle JM\,|S_z|\,JM\rangle = (g_J - 1)\,M \tag{21}$$

$$\langle JM\,|S_z|\,J+1, M\rangle = f(J, M) \tag{22}$$

$$\langle JM\,|S_x|\,J, M+1\rangle = \frac{g_J - 1}{2}\sqrt{(J-M)(J+M+1)} \tag{23}$$

$$\langle J+1, M\pm 1\,|S_x|\,J, M\rangle = \mp f(J, M)\,\frac{1}{2}\,\sqrt{(J\pm M+1)(J\pm M+2)/(J+M+1)(J-M+1)}$$

where (24)

$$f(J, M) = \left[\frac{(J+L+S+2)(-J+S+L)(J+S-L+1)(J+L-S+1)(J+M+1)(J-M+1)}{4\,(J+1)^2(2J+1)(2J+3)}\right]^{1/2}$$

Having established the general principles involved in evaluating the eigenvalues and eigenfunctions of the rare earth ions under the influence of pure crystal fields and the additional terms such as exchange and external fields (*33, 34*), we shall illustrate the calculation of magnetic, thermal and electrical properties in the next section.

B. *Interrelationship with Experiment*

1. *Magnetic Susceptibility*

Penney and *Schlapp* (*35*) many years ago discussed the influence of crystal field on the temperature dependence of magnetic susceptibility χ in paramagnetic materials. χ may be computed from the fundamental Van Vleck equation (*36*)

$$\chi = \frac{N_0}{H_{ext}} \frac{\Sigma \mu_i e^{-E_i/kT}}{\Sigma e^{-E_i/kT}} \tag{25}$$

where N_0 is the Avogadro number.

H_{ext} represents the external field and μ_i and E_i represent the magnetic moment and energy, respectively, of the ith crystal field level.

$$\mu_i = \langle \Gamma_i | \mu_{(Op)} | \Gamma_i \rangle \tag{26}$$

where $\mu_{(Op)} = gJ$ and Γ_i is the eigenfunction of the ith level. The magnetic moments of various rare earth ions under the influence of hexagonal crystal fields and external fields have been conveniently tabulated by *Segal* and *Wallace* (*25, 37*).

Penney and *Schlapp* (*35*) pointed out that for temperatures such that $kT > E_i$, Eq. (25) leads to a reciprocal susceptibility which is linearly dependent on T, thus satisfying Curie-Weiss law. At low temperatures significant deviations from the linear behavior may occur. In some instances where the interaction of magnetic ion with the crystal field is stronger than its interaction with the neighboring ions in the solid, the Van Vleck temperature-independent paramagnetism is observed. In these materials, the magnetic interaction is weak compared to crystal field interaction. In this case the magnetic entropy is lost on cooling toward O K through interactions with the crystal field. We will illustrate these effects with some specific examples in a later section of this article. Calculation of magnetic susceptibility using Eq. (25) for a known symmetry of the crystal and comparison with experimental results is one of the earliest recognized methods in the parametrization of crystal field intensities.

In the ordered materials, it is often necessary to calculate the magnetization using the crystal field theory. This may be done by diagonalizing the appropriate Hamiltonian and evaluating

$$M = N_0 \Sigma \mu_i e^{-E_i/kT} / \Sigma e^{-E_i/kT} \tag{27}$$

where M is the magnetization per gram atom of the rare earth ion. There are two cases which need special mention: those in which the exchange arises mainly due to the partner with which the rare earth is associated, such as RFe_2 or RCo_5 (see Sections IV C, D). In these cases the exchange due to R–R is assumed to be negligible. The second case is the one involving a nonmagnetic element as a partner to the rare earth ion such as RAl_2 (see Section IV A). In this instance it is necessary to evaluate the magnetization of the rare earth ion self-consistently.

2. *Heat Capacity and Free Energy*

The crystal field interaction has pronounced effects on the heat capacity behavior of the system. At very low temperatures, ions occupy the lowest crystal field states. With increasing temperature, excitation within the crystal field spectrum takes place resulting in a significant contribution to the heat capacity (*38*). This contribution is given by the expression

$$C_{CF} = \frac{R}{T^2} \frac{\partial^2 \ln Q}{\partial\left(\frac{1}{T}\right)^2} \tag{28}$$

where R is the gas constant and Q, the partition function ($= \Sigma\, e^{-E_i/kT}$). C_{CF} is also known as Schottky heat capacity. For computational convenience the above expression may be written as

$$C_{CF}/R = \frac{1}{Q}\left\{\Sigma\left(\frac{E_i}{kT}\right)^2 e^{-E_i/kT} - \left[\Sigma\left(\frac{E_i}{kT}\right) e^{-E_i/kT}\right]^2 / Q\right\} \tag{29}$$

At low temperatures (where $kT \ll E_2$), the above expression may be approximated by

$$C_{CF}/R = \frac{g_2}{g_1}\left(\frac{E_2}{kT}\right)^2 e^{-E_2/kT} \tag{30}$$

where g_1 and g_2 are the degeneracies of the first and second levels.

In the limit of low temperatures, it is easily seen that a plot of $\ln(C_{CF}\,T^2)$ vs T^{-1} gives a straight line, the slope of which is $-E_2/k$.

The free energy of the system at any temperature T under the influence of crystalline electric field may be calculated from the expression

$$A = -RT \ln Q \tag{31}$$

3. *Neutron Inelastic Scattering*

The techniques of optical absorption and fluorescence are not suitable in the study of metals and intermetallics because of the weak intensity of transitions and the inability of radiation to penetrate the metal. However, neutron inelastic scattering experiments have been used to provide quantitative information about the levels originating with the 4f electrons in these systems (*39*). In these experiments the sample is irradiated by a collimated monochromatic neutron beam and the scattered neutrons are analyzed according to the energy transfer

$$\hbar\omega = \frac{\hbar^2}{2m}(k_0^2 - k_1^2) \tag{32}$$

where m is the mass of the neutron, ω the frequency, k_0 and k_1 are the wave numbers of the incoming and outgoing neutrons. The corresponding momentum transfer is given by

$$\hbar\vec{Q} = \hbar(\vec{k}_0 - \vec{k}_1) \tag{33}$$

where $\vec{Q}$ is the scattering vector. In these experiments both energy gain and energy loss processes can be measured. This corresponds to the possibility of the rare earth ion either being de-excited from a higher to a lower state and the neutron gaining the corresponding energy or vice versa. Therefore one expects the measured energy spectrum to exhibit resonance peaks which can be related to the transitions between different crystal field levels. However, in the analysis it is very often found that a unique identification of the observed transitions is not possible on the basis of an energy level diagram alone; the relative intensities of the various transitions have to be considered. Details of experimental techniques and the problems associated with this method have been reviewed by *Furrer* (*40*). Inelastic neutron scattering is a powerful method for obtaining information about the crystal field interaction.

4. *Spin-Disorder Resistivity*

In rare earth metals and alloys, the exchange scattering of conduction electrons by localized moments gives rise to a contribution to the electrical resistivity known as spin-disorder resistivity. The original treatments (*41–44*) of this topic ignored the influence of crystalline electric fields on the spin-disorder resistivity. This is valid at temperatures sufficiently high in comparison to the crystal field splitting of the rare earth ion. In this "high temperature" approximation one obtains a temperature-independent expression for the spin-disorder resistivity that is proportional to the quantity $(g-1)^2 J(J+1)$. However, at low temperatures there is a very marked influence of the crystal field on spin-disorder resistivity as was first shown by *Rao* and *Wallace* (*45*).

The method proposed by *Rao* and *Wallace* (*45*) essentially takes into account (a) the Boltzmann distribution of rare earth ions among the crystal field states and (b) the contribution to the resistivity from inelastic scattering processes wherein the rare earth ion undergoes a transition from one crystal field state to another and the conduction electron gains or loses the corresponding amount of energy. Assuming an exchange interaction between the conduction electron s and rare earth spin $\vec{S}$ of the type

$$\mathscr{H} = -2\mathscr{J}_{sf}\, s \cdot \vec{S} \tag{34}$$

it was shown that the appropriate expression for spin-disorder resistivity $\rho_{s1}(T)$ in the first Born approximation may be written

$$\rho_{s1}(T) = \frac{3\pi Nm}{\hbar e^2 E_F} \mathscr{J}_{sf}^2 (g-1)^2$$

$$\times \sum_{m_s m_s' i i'} \langle m_s', i' | \vec{s} \cdot \vec{J} | m_s, i \rangle^2 \, p_i \, f_{ii'} \tag{35}$$

In the above expression N is the number of rare earth ions per unit volume, m_s and m_s' are the spins of the conduction electron states, i and i′ are the rare earth crystal field states, p_i is the Boltzmann probability that the rare earth ion is in the crystal field state of energy E_i, and $f_{ii'}$ are given by

$$f_{ii'} = \frac{2}{1 + \exp[(E_{i'} - E_i)/kT]} \tag{36}$$

It was shown that the above procedure gave a satisfactory description of the experimentally observed variation of spin disorder resistivity with temperature in $CeAl_2$ (*45*).

Evaluation of the spin-disorder resistivity in the second Born approximation, when J_{sf} is negative shows an upturn at low temperatures (Kondo effect). *Maranzana* (*46*) and later *Cornut* and *Coqblin* (*47*) showed that the anomalous temperature dependence of resistivity of compounds such as $CeAl_3$ could be explained satisfactorily by considering crystal field effects on Kondo resistivity.

5. *Magnetic Anisotropy*

This is one of the most interesting features of rare earth intermetallics since it underlies the interesting permanent magnet properties of the RCo_5 compounds, e.g., $SmCo_5$. In a substance which is to be used in the fabrication of permanent magnets there must be some powerful effect resisting the dealignment of the individual atomic moments against the influence of thermal agitation and magnetic domain formation. An external field is applied to sweep out the domain walls; some interaction must operate to maintain the situation when the external field is withdrawn. This interaction in the rare earth intermetallics is primarily the CFI.

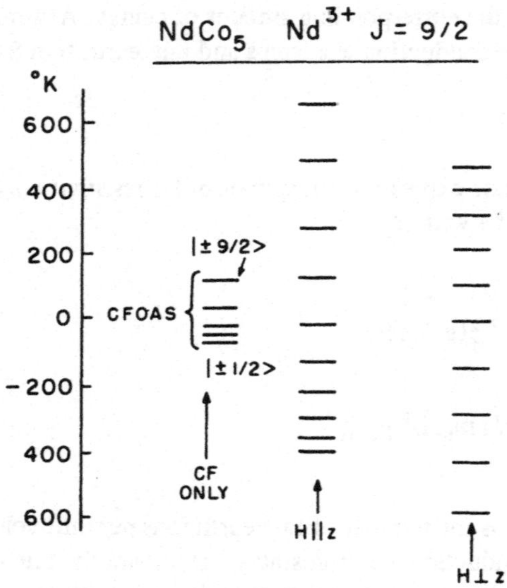

Fig. 2. The energy levels of Nd^{3+} ion under the influence of a second order crystal field and a magnetic field applied parallel and perpendicular to the crystallographic c-axis.

For Nd^{3+} ion, J = 9/2 and hence the ground state multiplet is tenfold degenerate. Under the influence of a crystal field of hexagonal symmetry, the degeneracy is partially removed. Five Kramer's doublets are formed with the eigenstates, $|\pm 1/2\rangle$, $|\pm 3/2\rangle$, $|\pm 5/2\rangle$, $|\pm 7/2\rangle$ and $|\pm 9/2\rangle$. The splitting of these levels under the influence of a second order crystal field is shown in Fig. 2. *Greedan* and *Rao* (*48*) considered the case of $NdCo_5$. Under the influence of the molecular field, the degeneracy is further lifted. The situation when the crystal field and magnetic interactions operate is shown in Fig. 2. The energy level structure is different depending upon the direction of the molecular field. It is clear that at low temperatures the free energy of the system would be lower with the molecular field (and moments) directed perpendicular to the hexagonal c-axis. Thus the CFI operates to maintain the alignment of moments in a certain direction relative to the crystallographic axes. The preferred direction may change with the alteration of temperature resulting in a so-called change of anisotropy. At this temperature the free energies of the two configurations are equal. Thus, study of magnetic anisotropy and its alteration with temperature may be used to provide information about the magnitude of the parameters B_n^m. This approach is illustrated later and in greater detail in the discussion of $HoAl_2$ and in the comments about the magnetic structures of the RNi_3 compounds.

IV. Discussion of Specific Families of Compounds

A. The Rare Earth-Aluminum Compounds

1. The RAl_2 Series

Rare earth-aluminum compounds of the composition RAl_2 have been extensively studied during the past two decades, both theoretically and experimentally (*49–84*). There are many reasons for the attention focused on RAl_2 compounds. These compounds are easy to prepare in single phase, and they possess the simple cubic Laves phase (C15) structure (*49*). The rare earth is the only magnetic species and therefore a comparative study of the series of RAl_2 compounds provides very useful information in regard to the crystal field interaction. Additionally, all the members of the family, including YAl_2, have been observed.

Early work on the structural aspects of RAl_2 compounds indicated that they crystallize in the cubic Laves phase structure with the rare earth ions experiencing cubic point symmetry (*49*). Low temperature magnetization studies by *Williams et al.* (*50*), *White et al.* (*51*) and *Mader* and *Wallace* (*52*) revealed that these compounds – except for YAl_2, $LuAl_2$ and $LaAl_2$, which contain no magnetic ions, and $EuAl_2$, which becomes antiferromagnetic at low temperatures – order ferromagnetically at low temperatures.[1]) Neutron diffraction studies on some RAl_2 compounds by *Nereson et al.* (*53*) confirmed the ferromagnetic nature of ordering.

Considerable work has been done recently to determine the details of the CFI in $ErAl_2$ compound. *Inoue et al.* (*54*) utilized calorimetric results to evaluate the CFI parameters in this material. Heat capacity measurements reported by *Inoue* (*55*) revealed that this material forms a cooperative magnetic phase at 10.2 K. From the earlier magnetization study it is clear that the material orders ferromagnetically. The contributions of the lattice and electronic terms to the heat capacity of $ErAl_2$ was estimated from a knowledge of the experimental heat capacities of $LaAl_2$ and $LuAl_2$ by an interpolation method. This additional heat capacity is plotted as ΔC_p in Fig. 3 as a function of temperature. It is clear that there is a significant contribution to the heat capacity due to the CFI, the maximum in the Schottky anomaly occurring at 23 K. These results are suggestive of the fact that the CFI is stronger than exchange interaction in $ErAl_2$.

The electronic ground state of Er^{3+} ion has a 16-fold degeneracy. The strong cubic crystalline electric field lifts this degeneracy into two doublets (Γ_6 and Γ_7)

---

[1]) $CeAl_2$ represents a special case. It may order antiferromagnetically but its behavior appears to be markedly affected by the Kondo interaction of the Ce^{3+} ion.

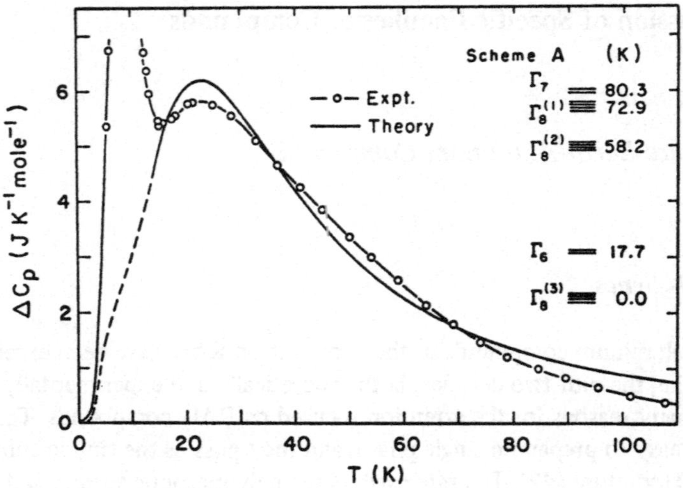

Fig. 3. Comparison of experimental excess heat capacity of $ErAl_2$ and the calculated curve based on the energy levels shown in the inset (Scheme *A*).

and three quartets ($\Gamma_8^{(1)}$, $\Gamma_8^{(2)}$ and $\Gamma_8^{(3)}$). The resulting energies of these levels vary as a function of x in a manner shown in Fig. 1. *Inoue et al.* (*54*) utilized the experimental ΔC_p curve to establish the crystal field overall splitting (CFOAS) and the relative magnitudes of the fourth- and sixth-order terms (as defined by parameter x) in the Hamiltonian [Eq. (7)]. Diagonalization of the Hamiltonian [Eq. (7)] of various values of x and W (which is proportional to the CFOAS) yields eigenvalues (E_i). These energy values are necessary to calculate the crystal field heat capacity using Eq. (29).

In this way *Inoue et al.* (*54*) obtained parameters x and W which best fit the experimental ΔC_p. The calculated Schottky heat capacity is compared with the experimental results in Figs. 3 and 4. Good agreement with experiment was obtained for two sets of parameters x and CFOAS: (1) $x = 0.5$, CFOAS = 80.3 K and (2) $x = -0.15$, CFOAS = 109 K. The Schottky heat capacity curves resulting from the calculations utilizing the above sets of parameters labeled Scheme *A* and Scheme *B*, respectively, are shown in Figs. 3 and 4.

As pointed out earlier in the Introduction, the crystal field parameters can also be calculated using the point charge model. This method can be applied to $ErAl_2$. In $ErAl_2$, the reference Er ion has 4 Er ions and 12 Al neighbors and the Er–Er and Er–Al distances are 3.38 Å and 2.24 Å, respectively. Calculations of B_4^0 and B_6^0 employing Eq. (4) result in the following:

$$B_4^0 = 2.89 \times 10^{-4} Z_1 - 3.48 \times 10^{-4} Z_2 \quad (37)$$

$$B_6^0 = -3.34 \times 10^{-7} Z_1 - 1.01 \times 10^{-7} Z_2 \quad (38)$$

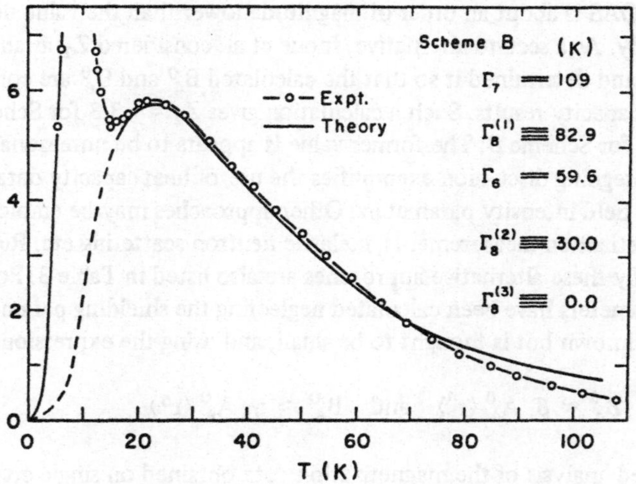

Fig. 4. Same as in Fig. 3 (Scheme *B*).

where Z_1 and Z_2 are the valences of the Er and Al ion, respectively, and B_4^0 and B_6^0 are in deg K.[2]) *Inoue et al.* (*54*) considered two alternatives in treating the above data. First, the valences Z_1 and Z_2 were taken as + 3 and the two parameters were evaluated. The results are summarized in Table 3. It is seen that the agreement between the experimentally determined and the calculated parameters using the point charge model is not satisfactory. The calculated crystal field intensity parameters and the

Table 3. Comparison of CFI parameters for $ErAl_2$ obtained from various experiments.

Quantity Measured		CFI Parameters		Ref.
		$B_4^0(10^{-3}$ K)	$B_6^0(10^{-5}$ K)	
Heat Capacity	Scheme *A*	− 1.77	− 0.76	(*54*)
	Scheme *B*	0.56	− 1.37	(*54*)
Magnetization		1.28	− 1.51	(*56, 57*)
Inelastic neutron scattering		1.86	− 0.70	(*58*)
PCM Calculations		− 0.21	− 0.15	(*54*)
Mössbauer		< 0	< 0	(*59*)
EPR		< 0 and	> 0	(*60*)
		> 0	> 0	(*60*)

[2]) Throughout this paper, the crystal field parameters B_n^m and $A_n^m \langle r^n \rangle$ are expressed in degree Kelvin.

related CFOAS is about an order of magnitude lower than the value determined experimentally. As a second alternative, Inoue et al. considered Z_2 as an adjustable parameter and determined it so that the calculated B_4^0 and B_6^0 are consistent with their heat capacity results. Such a calculation gives $Z_2 = +3.3$ for Scheme *A* and $Z_2 = +2.6$ for Scheme *B*. The former value is appears to be unreasonable.

The foregoing discussion exemplifies the use of heat capacity data to evaluate the crystal field intensity parameters. Other approaches may be employed – high field magnetization measurements, inelastic neutron scattering etc. Results obtained for $ErAl_2$ by these alternative approaches are also listed in Table 3. Point charge model parameters have been calculated neglecting the shielding parameter σ_n, which is not well known but is thought to be small, and using the expressions

$$B_4^0 \approx \beta_J A_4^0 \langle r^4 \rangle \quad \text{and} \quad B_6^0 \approx \gamma_J A_6^0 \langle r^6 \rangle \tag{39}$$

Detailed analysis of the magnetization data obtained on single crystals of $ErAl_2$ has been made by *Purwins et al.* (*56, 57*). This analysis confirms that Scheme *B* more aptly describes the CFI in $ErAl_2$, a conclusion which is strengthened by the observation that Scheme B also provides a better account of the heat capacity behavior of $ErAl_2$. The signs of B_4 and B_6 reported by *Heer et al.* (*58*) for polycrystalline $Er_cY_{1-c}Al_2$ by neutron spectroscopy lend additional support for the correctness of Scheme *B*. Mössbauer results of *Wiedemann* and *Zinn* (*59*) provide the signs for the CFI parameters which agree with Scheme *A*. Electron spin resonance results (*60*) on diluted $Er_{1-c}Y_cAl_2$ gave two sets of signs for the CFI parameters shown in Table 3.

In summary, we note that the set of parameters for the CFI obtained by *Inoue et al.* (*54*) (Scheme *B*) consistently accounts for not only the heat capacity behavior but also the magnetization results and the neutron scattering data. Calculations based on the point charge model give parameters which are nearly an order of magnitude smaller, and for B_4^0 lead to the wrong sign. The best crystal field overall splitting of the Er^{3+} ground J-manifold in $ErAl_2$ is ~ 109 K and the ground level is a $\Gamma_8^{(3)}$ quartet. The failure of the point charge model grows out of two unjustified simplifications: (1) neglect of non-Coulombic interactions and (2) neglect of screening by the conduction electrons. The point charge model is successful to the extent of correctly predicting that the magnitudes of B_4^0 and B_6^0 differ by roughly two orders of magnitude.

The CFI interaction of $HoAl_2$ has been recently examined in this laboratory (*61*) and by *Barbara et al.* (*62*). The interplay of the CFI and exchange interaction leads to some interesting properties in this compound. The heat capacity vs temperature curve by *Hill* (*63*) shown in Fig. 5 exhibits peaks at 20 K and 28 K. The latter is the temperature at which ferromagnetic ordering sets in. *Sankar et al.* (*61*) ascribed the origin of the anomaly at 20 K to a change in the easy direction of magnetization; for $T < 20$ K moments were presumed to lie in one crystallographic direction and for $T > 20$ K in another. Crystal field parameters were derived by calculating the free energy of the holmium ion under the influence of an exchange field self-consistently determined, and a crystalline electric field. Calculations were made with the exchange field operating along first the ⟨100⟩ direction, next along the ⟨110⟩ direction and

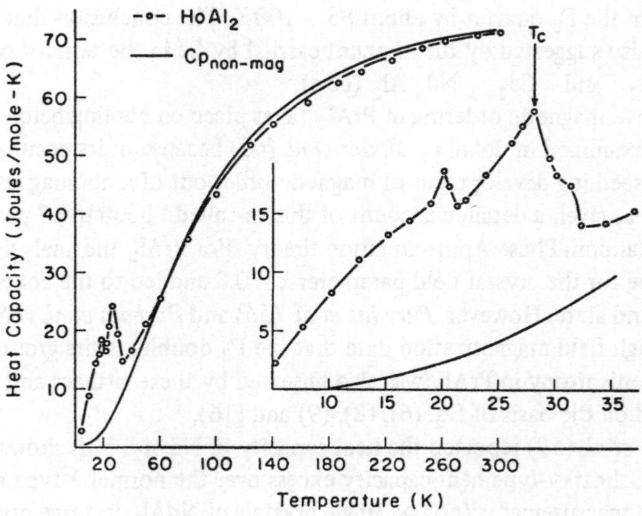

Fig. 5. Experimental heat capacity vs. temperature for $HoAl_2$. Heat capacity values of the non-magnetic part are also shown in the figure.

finally along the ⟨111⟩ direction. The calculations were made utilizing Eq. (6), (8) and (9). It was shown that only for certain ranges of parameters, viz. $A_4 \langle r^4 \rangle$ between -90 and $+90$ K and $A_6 \langle r^6 \rangle$ between 0 and -12 K, can a change in the direction of magnetization take place. With these parameters it was found that the moments are oriented in the ⟨110⟩ direction at low temperatures but with increasing temperature realign to the ⟨100⟩ direction. The validity of these calculations was independently confirmed by *Barbara et al.* (*62, 64*) using magnetization measurements on single crystals of $HoAl_2$ and by polarized neutron diffraction experiments. CFI parameters reported by these authors

$$A_4^0 \langle r^4 \rangle = 24\,\mathrm{K} \quad \text{and} \quad A_6^0 \langle r^6 \rangle = -5.6\,\mathrm{K}$$

also confirm the earlier studies (*61*). On the basis of these results an overall crystal field splitting of nearly 150 K is predicted for the ground J manifold with the lowest level being the triplet $\Gamma_5^{(1)}$.

Thus we have seen that in $ErAl_2$ and $HoAl_2$ the CFI plays a significant part in modifying the heat capacity behavior and in determining the magnetic anisotropy. This pattern of the influence of CFI appears to be present in the whole family of RAl_2 compounds. The effect of the CFI is clearly evident in $CeAl_2$ although as pointed out earlier, $CeAl_2$ appears to be a complicated example because of the influence of the Kondo interaction on the Ce^{3+} ion. Currently available information derived from analysis of the spin-disorder resistivity (*45*) and Schottky heat capacity excess (*65*) seems to suggest that the ground J = 5/2 manifold is split into a Γ_7 doublet separated from the Γ_8 quartet by about 85 – 100 K. The conclusion that the ground state is Γ_7 is also suggested by the moment carried by Ce in the ternary systems $Ce_{1-x}Pr_xAl_2$ and $Ce_{1-x}Nd_xAl_2$ *(65a)*.

Cooperative magnetic ordering in $PrAl_2$ takes place on cooling below 34 K. This material was examined in detail by *Mader et al.* *(66)* because of its peculiar feature involving the seeming development of magnetic order out of a nonmagnetic singlet. *Cooper* *(67)* has given a detailed account of this so-called "bootstrap" process employing the Random Phase Approximation theory. For $PrAl_2$ the analysis of *Mader et al.* *(66)* gave for the crystal field parameter $x = 0.8$ and led to the conclusion that Γ_1 is the ground state. However, *Purwins et al.* *(56)* and *Barbara et al.* *(68)* report on the basis of high field magnetization data that the Γ_3 doublet is the ground state. Considerable anisotropy in $PrAl_2$ was also observed by these authors and the results were analyzed on the basis of Eq. (6), (8), (9) and (16).

Deenadas et al. *(69)* reported the heat capacity of $NdAl_2$. This shows superimposition of a Schottky-type heat capacity excess over the normal λ-type anomaly. Magnetization measurements *(68)* on single crystals of $NdAl_2$ in three crystallographic directions in fields up to 350 kG and neutron inelastic scattering experiments *(70)* yielded consistent crystal field parameters ($A_4^0\langle r^4\rangle$ = 40 K and $A_6^0\langle r^6\rangle = -12$ K).

Additional evidence as to strong CFI in RAl_2 compounds stems from a study of the magnetization of $TbAl_2$. In this material a strong CFI and a powerful exchange field (which is proportional to Curie temperature $T_c = 105$ K) are present simultaneously. The crystal field interaction introduces a strong anisotropy in $TbAl_2$ which in turn modifies the temperature dependence of magnetization. At low temperatures magnetization is observed to follow the relationship $T^{3/2}\exp(-\Delta/kT)$ where Δ is the energy gap in the spin wave spectrum *(71)*. The absence of a strong CFI and the consequent absence of anisotropy would have resulted in a simple $T^{3/2}$ dependence of magnetization.

The crystal field parameters, the ground states, the easy directions of magnetization, T_c and CFOAS of the RAl_2 compounds are collected in Table 4. In summary,

Table 4. Best CFI parameters for RAl_2 compounds obtained from magnetic, thermal and electrical measurements *(45, 54, 61, 64, 65, 68, 69)*.

RAl_2	$A_4\langle r^4\rangle$ (K)	$A_6\langle r^6\rangle$ (K)	Ground State	μ_{sat} (μ_B)	Easy Direction	T_c (K)	CFOAS (K)
$CeAl_2$	45.7	–	Γ_7	0.7	–	–	85–100
$PrAl_2$	69.5	– 16.7	Γ_3	2.88	⟨100⟩	33	~ 200
$NdAl_2$	40	– 12.3	Γ_6	2.5	⟨100⟩	65	~ 200
$TbAl_2$	66.3	– 22.2	Γ_1	8.9	⟨111⟩	100	~ 150
$HoAl_2$	24	– 5.7	$\Gamma_5^{(1)}$	7.9	⟨110⟩[a]	32	~ 150
$ErAl_2$	26	– 8.4	$\Gamma_8^{(3)}$	7.8	⟨111⟩	14	109

[a]) At ~ 20 K the easy direction changes to ⟨100⟩.

we note that except for $GdAl_2$ the CFI significantly influences the behavior – magnetic moment, heat capacity, electrical resistivity and magnetic anisotropy – of the RAl_2 compounds. When the effect of CFI can be separated out it has proved possible to analyze the results so as to establish the specifics of the interaction – the magnitude of the fourth and sixth order interaction parameters, the nature of the ground state and the overall splitting. The latter values, again excepting $GdAl_2$, are in the range of 100 to 200 K.

2. *The RAl_3 Series*

Crystal field studies on RAl_3 compounds are complicated because of three changes in the equilibrium crystal structure across the series. Compounds to the left of $GdAl_3$ crystallize in the hexagonal Ni_3Sn structure and those to the right of $ErAl_3$ in the cubic Cu_3Au structure, the intermediate compounds possessing a mixture of the two types of atomic stacking in which the cubic character becomes more prominent for the smaller rare earth ions (*72, 73*).

To date only four compounds – $CeAl_3$, $PrAl_3$, $NdAl_3$ and $TmAl_3$ – have been thoroughly examined to obtain crystal field parameters (*74–84*). In this section we briefly point out some pertinent details.

The compound $PrAl_3$ provides a classic example where results obtained from heat capacity, susceptibility and spin-disorder resistivity provide a unique description of

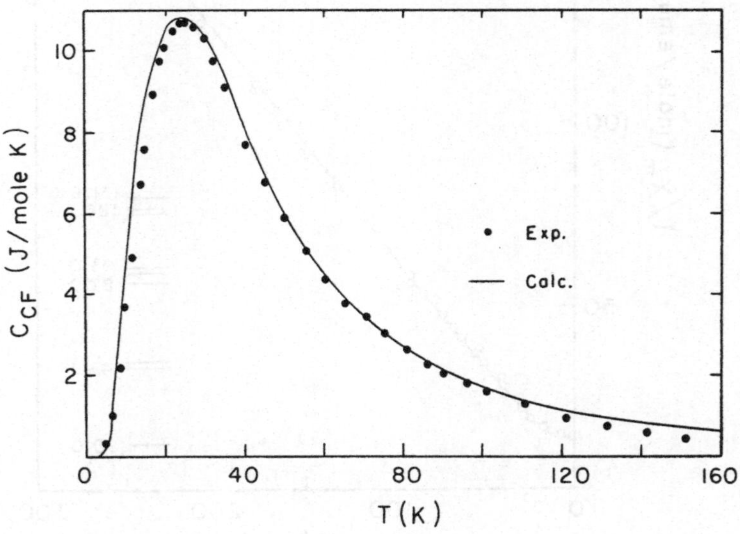

Fig. 6. Experimental and calculated crystal field heat capacity of $PrAl_3$ as a function of temperature according splitting scheme shown in Fig. 7.

the CFI of Pr^{3+}. *Mahoney et al.* (*74*) measured the heat capacities of $PrAl_3$ and $LaAl_3$ (employed as a "blank") in the temperature range 1.5 to 300 K. The excess heat capacity associated with the crystal field splitting of Pr^{3+} is indicated in Fig. 6 (*75*). The crystal field parameters that yield the best fit to the experimental heat capacity were found to be $B_2^0 = 2.74$ K, $B_4^0 = 0.041$ K and $B_6^0 = 0.0007$ K. This crystal field scheme also provided a satisfactory accounting of the susceptibility and spin-disorder resistivity of $PrAl_3$ as seen in Figs. 7 and 8. It is interesting to note that a point-charge calculation including only the contribution from the 12 nearest-neighbor Al ions, with $Z_{Al} = +3$, gave $B_2^0 = 3.27$ K, $B_4^0 = .059$ K and $B_6^0 = .0012$ K, parameters that are quite close to those provided by experiment. A notable feature is the dominance of the second order term over the others, a feature that is often revealed in hexagonal compounds. This, together with the point charge model, often enables one to make a "back-of-the-envelope" estimate of the CFI in a hexagonal compound if experimental data are lacking.

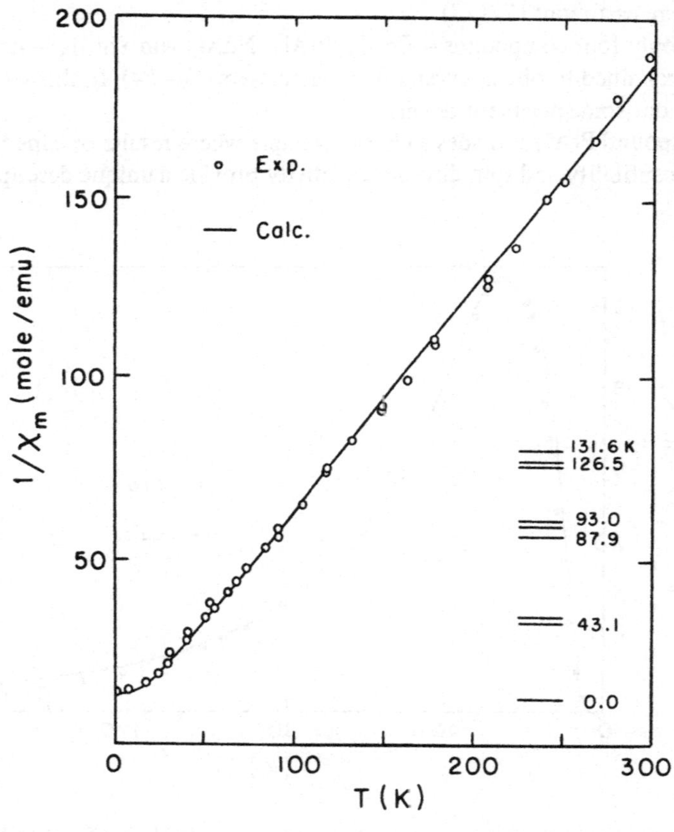

Fig. 7. Experimental and calculated inverse susceptibilities of $PrAl_3$ as a function of temperature.

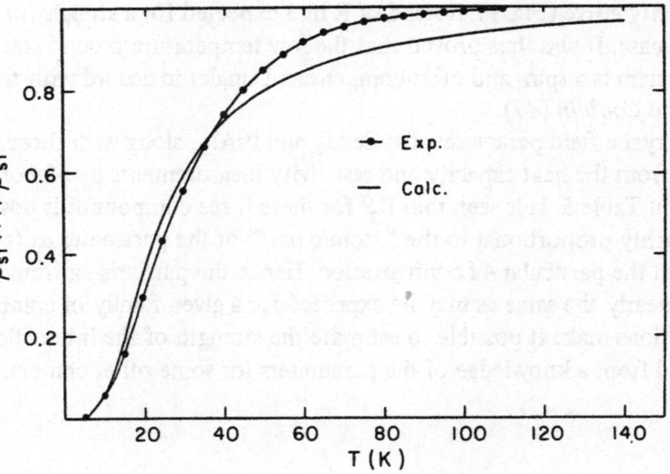

Fig. 8. Comparison of experimental and calculated resistivity ratio as a function of temperature for $PrAl_3$.

$CeAl_3$ is a very interesting case where the superposition of the Kondo effect on crystal field interaction gives rise to anomalous heat capacity (*76*) and resistivity behavior (*46, 77, 78*) at low temperatures. The heat capacity measurements of *Mahoney* (*75*) showed that the full entropy of R ln 6 was recovered under the excess

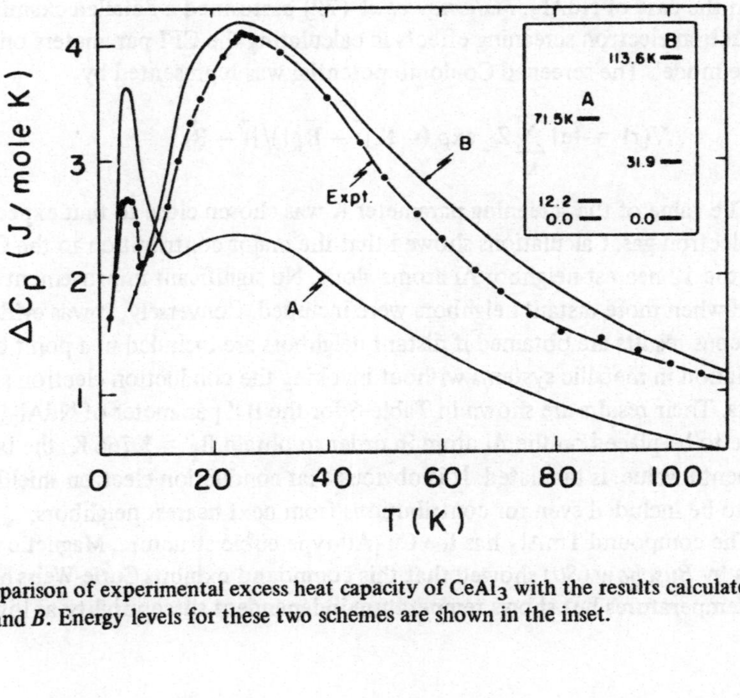

Fig. 9. Comparison of experimental excess heat capacity of $CeAl_3$ with the results calculated for Schemes *A* and *B*. Energy levels for these two schemes are shown in the inset.

heat capacity curve (Fig. 9), instead of R ln 3 expected for a straightforward crystal field-split case. It was thus proved that the low temperature ground state in this Kondo system is a spin- and orbit-compensated singlet in accord with the theory of *Cornut* and *Coqblin* (*47*).

The crystal field parameters for $CeAl_3$ and $PrAl_3$, along with those for $NdAl_3$ obtained from the heat capacity and resistivity measurements by *Mahoney et al.* (*79*) are given in Table 5. It is seen that B_2^0 for these three compounds is positive in sign and is roughly proportional to the "atomic part" of the parameter $\alpha_J \langle r^n \rangle$ which arises from the particular 4f configuration. Hence the part arising from the "environment" is nearly the same as may be expected for a given family of compounds. Such considerations make it possible to estimate the strength of the interaction in a certain compound from a knowledge of the parameters for some other compound of the family.

Table 5. Crystal field parameters (in deg K) for $CeAl_3$, $PrAl_3$ and $NdAl_3$ (*74, 76, 79*).

	$CeAl_3$	$PrAl_3$	$NdAl_3$
B_2^0	4.881	2.742	1.766
B_4^0	0.228	0.041	0.013
B_6^0		0.0007	0.0009

In the case of $NdAl_3$, *Mahoney et al.* (*79*) performed a detailed examination of conduction electron screening effects in calculating the CFI parameters on a point charge model. The screened Coulomb potential was represented by

$$V(r) = |e| \sum_k Z_k \exp(-K|\vec{r} - \vec{R}_k|)/|\vec{r} - \vec{R}_k| \qquad (40)$$

The value of the screening parameter K was chosen close to that expected for a free electron gas. Calculations showed that the major contribution to the CFI arose from the 12 nearest-neighbor Al atoms alone. No significant improvement was obtained when more distant neighbors were included. Conversely, it was evident that erroneous results are obtained if distant neighbors are included in a point charge calculation in metallic systems without invoking the conduction electron screening effects. Their results are shown in Table 6 for the B_2^0 parameter of $NdAl_3$. The charge to be placed on the Al atom in order to obtain $B_2^0 = 1.766$ K, the best experimental value, is tabulated. It is obvious that conduction electron shielding effects have to be included even for contributions from next nearest neighbors.

The compound $TmAl_3$ has the Cu_3Au-type cubic structure. Magnetic measurements by *Buschow* (*80*) showed that this compound exhibits Curie-Weiss behavior at high temperatures but shows temperature-independent susceptibility at low temper-

Table 6. Point charge calculations with and without conduction electron screening effects in $NdAl_3$ to obtain Z_{Al} that will yield $B_2^0 = 1.766$ K. Z_{Nd} is taken as + 3 and $\sigma_2 = 0.5$ (*79*).

	Nearest neighbors (12 Al ions)	Next nearest neighbors (12 Al + 8 Nd ions)	Shell of radius a (30 Al + 14 Nd ions)		Shell of radius 6a (8274 Al + 2902 Nd ions)	
	K = 0	K = 0	K = 0	K = 0.93 a.u.	K = 0	K = 0.93 a.u.
Z_{Al}	+ 4.16	− 10.06	− 4.58	+ 5.02	0.72	+ 4.26

atures, which is characteristic of a Van Vleck paramagnet. Analysis by *Buschow* (*80*) assuming a pure fourth order interaction and a very weak exchange field ($\theta_{ex} = -8$ K) gave an overall crystal field splitting of 435 K and led to a Γ_1 ground state. A more careful analysis of $TmAl_3$ was made by *Bucher et al.* (*81*) on the basis of heat capacity data. The analysis, which was similar to that by *Mahoney et al.* (*74*), described above, for $PrAl_3$, led to an overall splitting of 194 K and indicated Γ_2 to be the ground state. The importance of the sixth order interaction in this case is such that it cannot be safely neglected.

B. The Rare Earth-Nickel Compounds

1. The RNi_2 Series

RNi_2 compounds crystallize in the cubic Laves phase structure. Because of the simplicity of structure and the ease of preparation and characterization these materials have been extensively studied. From magnetic studies of this family of compounds, *Skrabek* and *Wallace* (*85*) established that nickel moment is zero in the ordered state and moreover that the moment of the rare earth atom is considerably reduced in comparison to that expected for a free trivalent rare earth ion. *Bleaney* (*86*) and *Skrabek* and *Wallace* (*85*) have interpreted this decrease in the saturation moment as arising from partial crystal field quenching of the orbital contribution to the total moment. In this respect the RNi_2 compounds behave like the RAl_2 compounds described in an earlier section above.

Bleaney (*86*) calculated the magnetic moments of the ground state levels of the rare earth ions in RNi_2 compounds assuming that only the fourth order crystal field interaction is important and using the crystal field parameters obtained from the point charge model [see Eq. (4)]. The nonmagnetic nature of Ni in the RNi_2 series indicates that nickel has acquired electrons from its rare earth partner in sufficient amount to fill its d-shell; consequently, the ion is in a $3d^{10}$ configuration and is uncharged. Bleaney in his treatment of the RNi_2 series assumed zero charge on Ni and regarded the rare earth ion as tripositive. Subject to these assumptions and working within the point charge model he derived the results shown in Fig. 10 and included in Table 7. The experimentally obtained saturation moments are seen to be substantially less than gJ, the moment expected for the free ion. This indicates that the moment has been reduced by the influence of the crystal field.

If the ground state, which was identified in each case by Bleaney, is well removed from all other crystal field states, the measured moment will give the moment of the ground state. If it is not well removed and if there is an exchange or an applied field, the excited states will make an appreciable contribution to the measured moment in two ways: First and simplest, because of the Boltzmann distribution excited states will be populated to some extent and these will contribute to the measured moment; this effect occurs even if there is no exchange or applied field. Second, the applied field or the exchange field will modify the ground state by admixture of the excited states. For example, with Nd^{3+} in $NdNi_2$ when the ion is exposed to a magnetic field the Γ_6 ground state will be admixed to some extent with the excited Γ_8 states. The requirements of the Boltzmann distribution and the admixture effect will both be reflected in the measured moment if the ground state is not well removed from the excited states, whereas if the excited states are energetically distant from the ground state (compared to $k\,T_m$, where T_m is the temperature of the measurement), the measured moment will be that for the ground state.

Table 7. Characteristics of RNi_2 indicating the influence of the Crystal Field Interaction (*86, 104*).

Compound	Ground state	Multiplicity	Calculated moment: Simple ground state	Calculated moment: Compound ground state	gJ	Exptal saturation moment μ_B	First excited state (K) calculated
$CeNi_2$ [a])	Γ_7	2	0.71	2.14	2.14	–	284
$PrNi_2$	Γ_1	1	0	2.07	3.2	0.86 [b])	77
$NdNi_2$	Γ_6	2	1.33	3.25	3.28	1.80	47
$SmNi_2$ [c])	Γ_7	2	0.24	0.71	0.71	0.25	112
$TbNi_2$	Γ_1	1	0	5.62	9.0	7.8	27
$DyNi_2$	Γ_6	2	3.35	8.02	10.0	9.2	12
$HoNi_2$	Γ_1	1	0	9.96	10.0	8.4	10
$ErNi_2$	$\Gamma_8^{(3)}$	4	5.50	6.23	9.0	6.8	36
$TmNi_2$	Γ_1	1	0	4.37	7.0	3.2 [b])	37
$YbNi_2$	Γ_6	2	1.33	4.0	4.0	–	155

a) Results are for tripositive rare earth ions. Ce in $CeNi_2$ appears to be quadripositive and hence is nonmagnetic.

b) This is not the saturation moment. It is the measured moment at 2 K and 20 kOe.

c) J-mixing is important in systems containing Sm^{3+}.

The example of $NdNi_2$ is instructive. The 4.2 K measured Nd^{3+} moment is 1.62 μ_B. The free ion moment is 3.28 μ_B. Clearly the CFI has effected a twofold reduction in the moment. The moment of the Γ_6 state is 1.33 μ_B. This was worked out many years ago by *Penney* and *Schlapp* (*35*) and is readily derivable from the eigenfunctions obtained for Nd^{3+} in a cubic field. Actually 1.33 μ_B is the so-called permanent moment of the Nd^{3+} ion in a cubic field; it is the moment in the absence of a magnetic field. If a magnetic field is applied, the admixture of the Γ_8 states adds an "induced component" to the Γ_6 moment. Hence the moment is 1.33 $\mu_B + \alpha H$, where α is the magnetic polarizability of the Γ_6 state and H is the magnetic field strength. *Penney* and *Schlapp* (*35*) showed that α varies inversely with the intensity of the CFI. *Wallace et al.* (*65*) found the overall crystal field splitting for Nd in $NdNi_2$ to be 430 K. Using this and assuming a pure fourth order interaction, which was the basis of the Penney-Schlapp calculation, $\alpha = 2.4 \times 10^{-3}$ μ_B/kOe. The effective magnetic field in $NdNi_2$, estimated from its Curie temperature of 20 K, is 113 kOe. Thus the induced component is 0.27 μ_B and the total moment of the ground state is 1.60 μ_B which is in excellent agreement with experiment.

Heat capacity work on $NdNi_2$ is consistent with the inferences drawn from the magnetic studies (*65*). Molar heat capacities of $NdNi_2$ show a λ-type excess heat capacity peaking at ~ 10 K. The difference between the $NdNi_2$ and its non-magnetic counterpart Ce^{4+} Ni_2 heat capacities is Schottky-type peaking at ~ 50 K. Referring to Fig. 10, we consider the λ-type excess to be due to the destruction of ferromagnetic order in the ground state Kramer's doublet and the Schottky thermal anomaly to

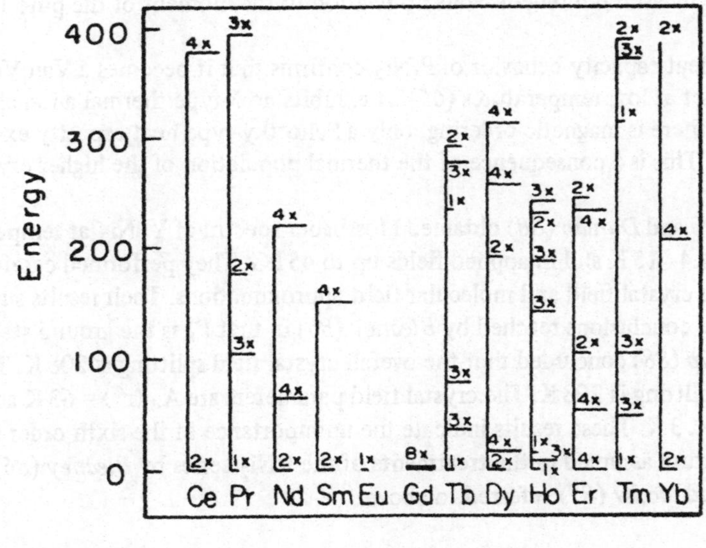

Fig. 10. Splitting pattern for R^{3+} ions in a pure fourth order cubic crystal field. The energy is given in arbitrary scale; the degeneracies of the levels are indicated.

originate with thermal population of the excited crystal field states. We therefore expect the ΔS associated with the magnetic disordering within the ground state to be $R \ln 2$. Experiment gives for this entropy change $1.06 R \ln 2$.

The ground states of Pr, Tb, Ho and Tm in the RNi_2 series are singlets. These singlet states do not carry a moment. The reason for this can be easily seen taking Pr^{3+} as an example

$$\Gamma_1 = a_1 |-4\rangle + a_2 |0\rangle + a_3 |4\rangle, \; a_1 = a_3 \neq a_2 \tag{41}$$

Applying the magnetic moment operator to this eigenfunction, one obtains for the moment

$$\mu = gJ \sum_i a_i^2 M \tag{42}$$

Thus in this case $\mu = 0$. If exchange is sufficiently strong to admix significant amounts of the excited states with the ground state, the above relationship between a_1, a_2 and a_3 no longer holds in the sense that a_1 and a_3 are no longer equal. Hence μ no longer vanishes. Careful work by *Wallace* and *Mader* (*87*) showed that $PrNi_2$ is a Van Vleck paramagnet. This indicates that the lowest lying level is the pure (or essentially pure) Γ_1 state. Exchange in $PrNi_2$ is very weak and hence the admixture of the higher states is negligible. In addition the crystal field splitting is sufficiently large that thermal population of the higher crystal field states is negligible. The situation in $TmNi_2$ may be similar but is less clearcut. In $TbNi_2$ and $HoNi_2$ thermal population of the higher (moment-carrying) states and admixture is extensive and hence the measured moment bears no relation to the moment of the pure Γ_1 ground state.

The heat capacity behavior of $PrNi_2$ confirms that it becomes a Van Vleck paramagnet at low temperatures (*65*). It exhibits no λ-type thermal anomaly expected if there is magnetic ordering; only a Schottky-type heat capacity excess is observed. This is a consequence of the thermal population of the higher crystal field states.

Nowik and *Dunlap* (*88*) obtained Mössbauer spectra of $YbNi_2$ at temperatures between 1.4–15 K and in applied fields up to 45 kG. They performed careful analysis within the crystal field and molecular field approximations. Their results support the qualitative conclusions reached by *Bleaney* (*86*) in that Γ_6 is the ground state. *Nowik* and *Dunlap* (*88*) concluded that the overall crystal field splitting is 208 K. The crystal field splitting is 208 K. The crystal field parameters are $A_4 \langle r^4 \rangle = 63$ K and $|A_6 \langle r^6 \rangle| < 3$ K. These results indicate the unimportance of the sixth order interaction, a feature assumed in the treatments of the RNi_2 series by *Bleaney* (*86*) and *Wallace* and *Mader* (*87*) referred to above.

2. The RNi_3 Series

The RNi_3 compounds crystallize in the $PuNi_3$-type hexagonal structure (see Fig. 11). The unit cell for this structure is complex, containing 36 atoms. There are two inequivalent sites (which we term I and II) occupied by the rare earth (*89*). The RNi_3 compounds order at temperatures of 116 K (for $GdNi_3$) and below (*90*). The interplay of exchange and the CFI alluded to in section III B 5 above is evident in their structural features, specifically the alignment of magnetic moments with respect to the crystallographic directions.

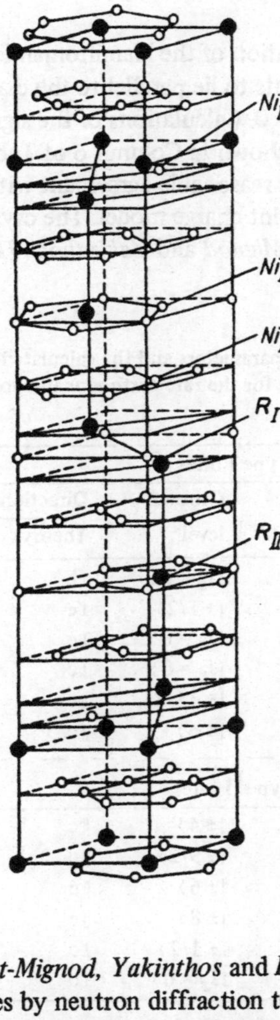

Fig. 11. Crystal structure of RNi_3.

Rossat-Mignod, Yakinthos and *Paccard* (*91–93*) studied several members of the RNi_3 series by neutron diffraction techniques and obtained magnetic structure information. They interpreted their results using the point charge model. Point symmetries of R(I) and R(II) ions are D_{3d} and C_{3v}.

$$\mathscr{H}_{CF}^{(I)} = B_2^0 O_2^0 + B_4^0 O_4^0 + B_4^3 O_4^{3(s)} + B_6^0 O_6^0 + B_6^3 O_6^{3(s)} + B_6^6 O_6^6 \quad (43)$$

$$\mathscr{H}_{CF}^{(II)} = B_2^0 O_2^0 + B_4^0 O_4^0 + B_4^3 O_4^{3(c)} + B_6^0 O_6^0 + B_6^3 O_6^{3(c)} + B_6^6 O_6^6 . \quad (44)$$

Rossat-Mignod and *Yakinthos* (*91*) retained only the leading term $B_2^0 O_2^0$ in each case since experience indicates that the higher order terms are small. Hence,

$$\begin{aligned}\mathscr{H}_{CF}^{(I)} &= B_2^0 (I)\, O_2^0 \\ &= A_2^0 (I) \langle r^2 \rangle (1 - \sigma_2)\, \alpha_J\, [3 J_z^2 - J(J+1)]\end{aligned} \quad (45)$$

with a similar expression for $\mathscr{H}_{CF}^{(II)}$. Diagonalization of the Hamiltonian leads to the result that the low energy situation is for moments to lie parallel to the c-axis for $B_2^0 < 0$ and perpendicular to the c-axis for $B_2^0 > 0$. Calculations of the sign of B_2^0 led *Rossat-Mignod* and *Yakinthos* to the results shown in Column 6 of Table 8. Comparison with experiment (Column 7) reveals reasonable agreement with crystal field theory considering the limitations in the point charge model. The crystal field splitting pattern for $ErNi_3$ calculated by *Rossat-Mignod* and *Yakinthos* (*91*) is shown in Fig. 12.

Table 8. Point charge model crystal field parameters and the calculated and the experimental magnetic moment directions for the rare earth ions in two types of sites in RNi_3 compounds (*91*).

	B_2^0 (K)	Δ[a] (K)	δ[b] (K)	Ground level	Direction of moment: Theory	Direction of moment: Expt.
Type I Sites						
$PrNi_3$	− 1224	1238	86	$\lvert J_z = 0\rangle$	⊥c	∥c
$NdNi_3$	− 1166	446	36	$\lvert \pm 1/2\rangle$	⊥c	30°
$TbNi_3$	− 920	1000	29	$\lvert J_z = 0\rangle$	⊥c	20°
$HoNi_3$	− 893	390	6	$\lvert J_z = 0\rangle$	⊥c	⊥c
$ErNi_3$	− 835	390	86	$\lvert \pm 15/2\rangle$	∥c	80°
$TmNi_3$	− 706	763	230	$\lvert \pm 6\rangle$	∥c	∥c
Type II Sites						
$PrNi_3$	533	533	230	$\lvert \pm 4\rangle$	∥c	∥c
$NdNi_3$	490	190	72	$\lvert \pm 9/2\rangle$	∥c	20°
$TbNi_3$	375	400	130	$\lvert \pm 6\rangle$	∥c	35°
$HoNi_3$	330	130	29	$\lvert \pm 8\rangle$	∥c	90°
$ErNi_3$	317	130	4	$\lvert \pm 1/2\rangle$	⊥c	60°
$TmNi_3$	280	300	10	$\lvert J_z = 0\rangle$	⊥c	∥c

[a]) Splitting of the ground J multiplet.
[b]) Splitting of the first excited level from the ground level.

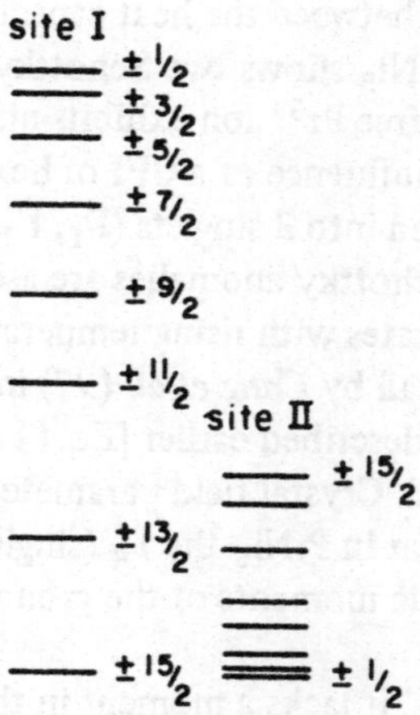

Fig. 12. Splitting pattern for two types of Er^{3+} ions in $ErNi_3$.

3. *The RNi_5 Series*

RNi_5 compounds crystallize in $CaCu_5$-type hexagonal crystal structure (*94*). Magnetic and thermal properties reported by *Nesbitt et al.* (*95*), *Wallace* and *Aoyagi* (*96*), *Craig et al.* (*97*), *Marzouk et al.* (*98*) and *Sankar et al.* (*99*), revealed the importance of crystal field interactions in these compounds. As an illustration, we will describe the magnetic thermal properties of $PrNi_5$, which exhibits particularly interesting crystal field features (*97*).

The magnetization of $PrNi_5$ exhibits a maximum at about 13 K (see Fig. 13). This was originally interpreted as arising from cooperative antiferromagnetic ordering; however, later studies of heat capacity behavior of this compound failed to show a λ-type anomaly in the neighborhood of 13 K, indicating that the maximum in λ did

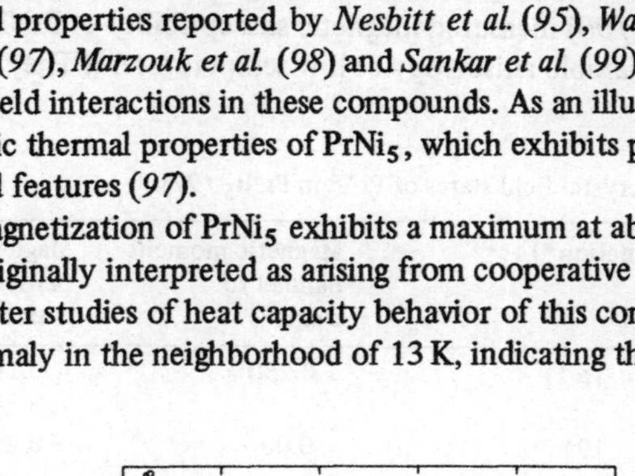

Fig. 13. Variation of magnetization of polycrystalline $PrNi_5$ in an external field of 19 kOe as a function of temperature.

not accord with a 13 K Néel point. The difference between the heat capacity of $PrNi_5$ and that of its non-magnetic counterpart $LaNi_5$ shows two Schottky-type thermal anomalies peaking at ~ 20 and 50 K. The free Pr^{3+} ion exhibits ninefold degeneracy (J = 4) in the ground state. Under the influence of a CFI of hexagonal symmetry the ground state multiplet is decomposed into 3 singlets (Γ_1, Γ_3 and Γ_4) and 3 doublets (2 Γ_5 states and a Γ_6 state). The Schottky anomalies are ascribed to thermal repopulation of the excited crystal field states with rising temperature.

The heat capacity results were analyzed in detail by *Craig et al.* (*97*) in terms of crystal field interaction employing a Hamiltonian described earlier [Eq. (11)]. An overall crystal field splitting of 250 K was observed. Crystal field parameters which best describe the experimental results indicated that in $PrNi_5$ the Γ_4 (singlet) is the ground state. Energies, eigenfunctions and magnetic moments of the ground and excited crystal field states are shown in Table 9.

Since the ground state is non-magnetic (that is, it lacks a moment in the absence of exchange or an applied field), $PrNi_5$ is a Van Vleck paramagnet. The lack of a magnetic moment in the ground state prevents it from displaying cooperative magnetic ordering at low temperatures. The maximum in susceptibility at ~ 13 K is caused by thermal excitation into the low-lying Γ_5 state (38 K above the ground state). The occurrence of this maximum is unusual. It is a peculiarity of the system brought on by the particular features of the crystal field splitting.

Recently *Andres et al.* (*100*) measured magnetic susceptibility of $PrNi_5$ single crystals and observed considerable anisotropy at low temperature.[3] They observed

Table 9. Characteristics of the crystal field states of Pr^{+3} in $PrNi_5$ (*97*).

State	Energy (°K)[a]	Eigenfunction[a]	Magnetic moment parallel to c-axis[b]	Magnetic moment perpendicular to c-axis[b]
Γ_6	99.5	$\vert\pm 1\rangle$	± 0.250	– 0.1557 – 0.3927
Γ_1	79.6	$\vert 0\rangle$	0.00	+ 0.2856
Γ_5	79.2	$0.178\vert\mp 4\rangle + 0.984\vert\pm 2\rangle$	+ 0.445 – 0.454	+ 0.1243 + 0.1239
Γ_3	– 61.5	$0.707[\vert 3\rangle + \vert -3\rangle]$	– 0.073	+ 0.0066
Γ_5	– 112.6	$0.984\vert\mp 4\rangle - 0.178\vert\pm 2\rangle$	– 0.949 + 0.955	+ 0.0043 – 0.0381
Γ_4	– 150.5	$0.707[\vert -3\rangle - \vert 3\rangle]$	+ 0.073	+ 0.0507

[a]) In zero magnetic field.
[b]) In a magnetic field of 20 kOe. The quantities tabulated are given as a fraction of gJ, the free ion moment in Bohr magnetons.

[3]) The authors are grateful to *Dr. Andres* of the Bell Laboratories for communicating his results prior to publication.

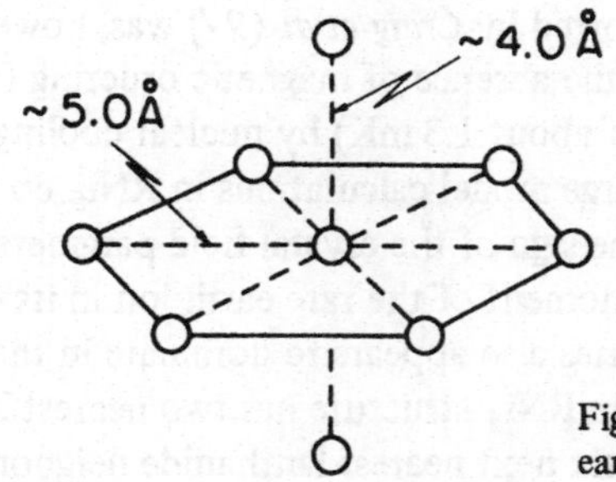

Fig. 14. Crystal environment showing only rare earth neighbors to the reference R ion in RNi_5.

that the susceptibility exhibits a maximum at nearly 15 K when a field is applied perpendicular to the c-axis, whereas χ measured with the field along the c-axis was found to be normal. These results were interpreted on the basis of the fact that the ratio B_6^6/B_6^0 in Eq. (10) deviates considerably from ideal behavior. The presence of

Table 10. Comparison of experimental and calculated magnetic moments and anisotropies in RNi_5 compounds (*95, 96, 101*).

Data for Lowest States of 4f Ions in a Crystal Field $\mathscr{H} = B_2^0 O_2^0$.

n = No. of 4f electrons	4, 5, 11, 12, 13	1, 2, 3, 8, 9, 10	
Sign of B_2^0	negative	positive	
Orientation of moment	parallel to hexagonal axis	perpendicular to hexagonal axis	
	n odd or even	*n* even	*n* odd
Ground state	$\|\pm J\rangle$	$\|0\rangle$	$\|\pm 1/2\rangle$
First excited state	$\|\pm(J-1)\rangle$	$\|\pm 1\rangle$	$\|\pm 3/2\rangle$
Minimum moment	gJ	0	$\frac{1}{2}g(J+\frac{1}{2})$
"Maximum" moment	gJ	$g[\frac{1}{2}J(J+1)]^{1/2}$	$\frac{1}{4}g[J+\frac{1}{2}+(5J^2+5J-11/4)^{1/2}]$

Ln	Minimum	"Maximum"	Experiment (Saturation)	Orientation
Ce	1.29	1.56	–	perpendicular
Pr	0	2.53	0.4	perpendicular
Nd	1.82	2.91	2.25	perpendicular
Sm	0.71	0.71	0.7	parallel
Gd	7.0	7.0	6.8	–
Tb	0	6.88	7.1	perpendicular
Dy	5.33	8.60	9.05	perpendicular
Ho	0	7.50	8.1	perpendicular
Er	9.0	9.0	7.8	parallel
Tm	7.0	7.0	6.7	parallel
Yb	4.0	4.0	3.6	parallel

a non-magnetic singlet ground state reported by *Craig et al.* (*97*) was, however, confirmed by *Andres et al.* (*100*). Further, the absence of magnetic ordering facilitated production of very low temperatures (of about 1.3 mK) by nuclear cooling.

Bleaney (*101*) performed point charge model calculations in RNi_5 compounds and established a correlation between the sign of the crystal field parameter and the directional preference of the magnetic moment of the rare earth ion in its ordered state. The second order term in RNi_5 series also appears to dominate in the crystal field Hamiltonian. Each lanthanide ion in RNi_5 structure has two nearest lanthanide neighbors along the hexagonal axis and six next nearest lanthanide neigbors in the plane normal to the hexagonal axis, as shown in Fig. 14. Nickel in this compound is non-magnetic and therefore its effect is neglected in these calculations. Bleaney's results correlating the sign of crystal field parameter and moment direction are summarized in Table 10. It is noted that the agreement between theory and experiment seems to be satisfactory.

C. The Rare Earth-Cobalt Compounds

1. The RCo_2 Series

The RCo_2 compounds crystallize in the cubic Laves phase structure (*102*). Magnetic properties of these compounds have been reported by *Wallace et al.* (*103, 104*), *Crangle* and *Ross* (*105*) and *Lemaire* (*106*). Results obtained have been described in detail by *Wallace* (*107*). The CFI appears to be a strong factor in modifying several properties in this series of compounds. In this respect the behavior is similar to that described earlier in connection with the RAl_2 and RNi_2 compounds, the difference, however, being that in RCo_2 compounds both R and Co ions carry a magnetic moment (*108, 109*). The compound $NdCo_2$ has been examined in great detail, both theoretically and experimentally with an objective to establish the role of the CFI (*110–113*). We choose this example to illustrate the influence of the CFI in RCo_2 compounds.

The neutron diffraction experiments of *Moon et al.* (*109*) and the analysis of the low temperature specific heat data of da *Silva et al.* (*110*) showed that the saturation moment of the Nd^{3+} ion in $NdCo_2$ is 2.6 μ_B. This is considerably smaller than the free ion value of 3.27 μ_B. This difference has been attributed to the partial quenching of the magnetic moment induced by the CFI. The Curie temperature of $NdCo_2$ is 98 K. Heat capacity measurements by *Thompson* (*111*) revealed two heat capacity anomalies. One was clearly associated with ferromagnetic ordering; an additional anomaly in the heat capacity was observed at 45 K, as shown in Fig. 15. The magnetic studies reported earlier by *Farrell* and *Wallace* (*104*) also showed an anomalous behavior at 45 K. *Thompson* (*111*) first predicted the possibility of a spin reorientation of the rare earth ion as a consequence of the combined influence of a strong CFI and an exchange interaction. Recently *Dublon et al.* (*112*) performed magnetization meas-

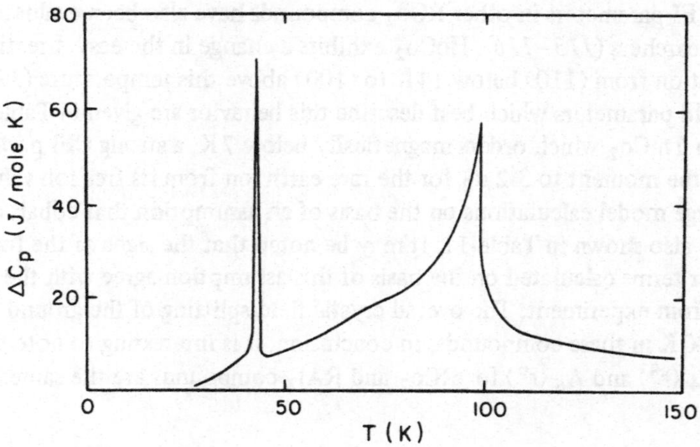

Fig. 15. Excess heat capacity of $NdCo_2$ as a function of temperature.

urements in low applied fields on $NdCo_2$ and also obtained additional data from Mössbauer study of Fe^{57} doped in $NdCo_2$. Their experiments confirmed the existence of a crystal field induced spin reorientation from ⟨110⟩ direction below 42 K to ⟨100⟩ direction above 42 K. The CFI parameters which best describe the above results have been evaluated by them to be

$$25 < A_4 < 40\,\mathrm{K}\,a_0^{-4} \quad \text{and} \quad -0.088 < \frac{A_6}{A_4} < -0.068\,a_0^{-2}\,.$$

Calculations performed by *Gignoux et al.* (*113*) also confirm the above results.

Table 11. Crystal field parameters in deg K obtained from experiment and calculated on the basis of PCM for RCo_2 compounds (*112–116*).

	$NdCo_2$	$HoCo_2$	$TmCo_2$
$A_4 \langle r^4 \rangle$			
EXPT	60–100	140	160
PCM	30	33	34
$A_6 \langle r^6 \rangle$			
EXPT	– 20 to – 40	– 18	– 10
PCM	– 0.89	– 1	– 1
Easy direction of magnetization at 4.2 K	⟨110⟩ [a]	⟨110⟩ [b]	⟨111⟩

[a]) Changes to ⟨100⟩ above 42 K.
[b]) Changes to ⟨100⟩ above 14 K.

The CFI parameters in other RCo_2 compounds have also been evaluated by several researchers (*113–116*). $HoCo_2$ exhibits a change in the easy direction of magnetization from ⟨110⟩ below 14 K to ⟨100⟩ above this temperature (*114*). The crystal field parameters which best describe this behavior are given in Table 11 for $HoCo_2$. In $TmCo_2$ which orders magnetically below 7 K, a strong CFI partially quenches the moment to 3.2 μ_B for the rare earth ion from its free ion value of 7 μ_B. Point charge model calculations on the basis of an assumption that cobalt carries no charge are also shown in Table 11. It may be noted that the signs of the fourth and sixth order terms calculated on the basis of this assumption agree with the parameters deduced from experiment. The overall crystal field splitting of the ground multiplet is 250–300 K in these compounds. In conclusion, it is interesting to note that the signs of $A_4 \langle r^4 \rangle$ and $A_6 \langle r^6 \rangle$ for RCo_2 and RAl_2 compounds are the same.

2. *The RCo_5 Series*

As pointed out in section III B 5, the strong CFI of the rare earth ion is responsible for the powerful magnetocrystalline anisotropy exhibited by certain of the RCo_5 systems (*117–124*). *Greedan* and *Rao* (*48*) analyzed the behavior of the rare earth ion anisotropy on the basis of a model in which the rare earth ion is simultaneously subjected to a crystalline electric field and an exchange (or molecular) field. Point charge model estimates were used to establish the preponderance of the second order term in the crystal field Hamiltonian. They drew attention to an interesting systematization: the sign of B_2^0 determines the anisotropy at low temperatures. For example, if B_2^0 is negative, the rare earth moment orients along the axial direction and if B_2^0 is positive, the rare earth moment lies either in the basal plane or a cone. These results are clearly illustrated in Fig. 16 and Table 12. *Greedan* and *Rao* (*48*) calculated the temperature variation of the free energy for the rare earth ions in RCo_5 under the influence of CFI and exchange interactions in the two directions (J_x and J_z) and compared these results with experiment (*117–119*). They showed that the CFI parameters listed in Table 12 correctly predict the temperature variation of anisotropy as shown in Fig. 17. These results are also in agreement with the heat capacity behavior of some RCo_5 compounds (*124*). As an illustration, the magnetic heat capacity behavior of $NdCo_5$ due to the CFI is shown in Fig. 18. The CFI parameters deduced by *Keller et al.* (*124*) confirm the calculations of *Greedan* and *Rao* (*48*).

The behavior of $SmCo_5$ and the influence of CFI on the anisotropy of the rare earth sublattice are discussed in section IV F below.

Table 12. Crystal Field Parameters and Directional Preference for RCo_5 Compounds-$CaCu_5$ Structure (*48, 117–119, 122*).

R	$g\mu_B H_{mol}$ (K)	$A_2^0\langle r^2\rangle$ (K)	B_2^0 (K)	Rare Earth Sublattice Preferred Direction	
				(Theory)	(Expt.)
Pr	84	– 1110	+ 4.5	plane	cone
Nd	115	– 1030	+ 1.3	plane	plane
Sm	–	– 970	– 7.9	axis	axis
Tb	210	– 775	+ 1.4	plane	plane
Dy	140	– 745	+ 0.9	plane	plane
Ho	104	– 720	+ 0.3	plane	plane
Er	84	– 680	– 0.4	axis	axis

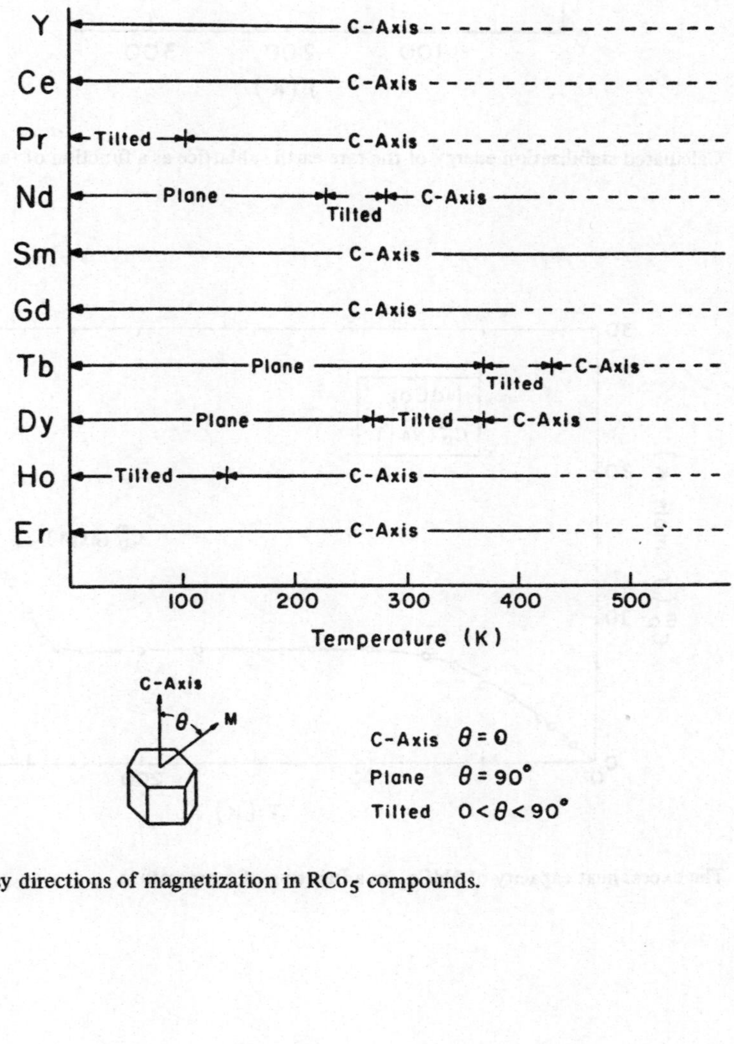

Fig. 16. Easy directions of magnetization in RCo_5 compounds.

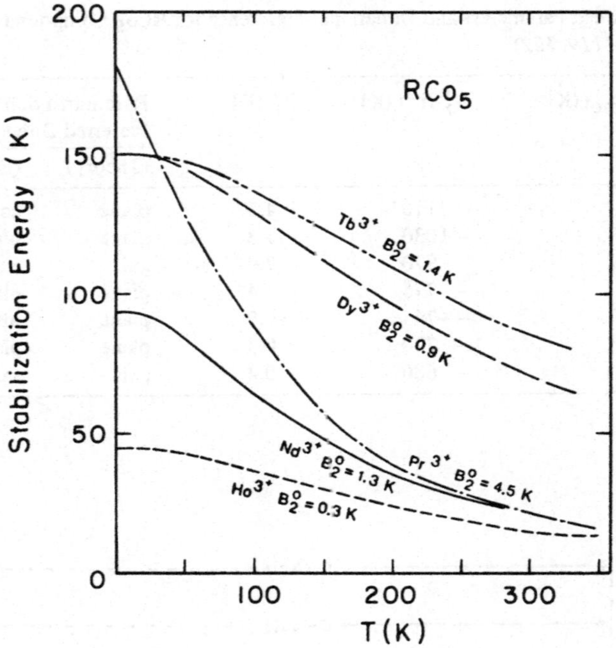

Fig. 17. Calculated stabilization energy of the rare earth sublattice as a function of temperature.

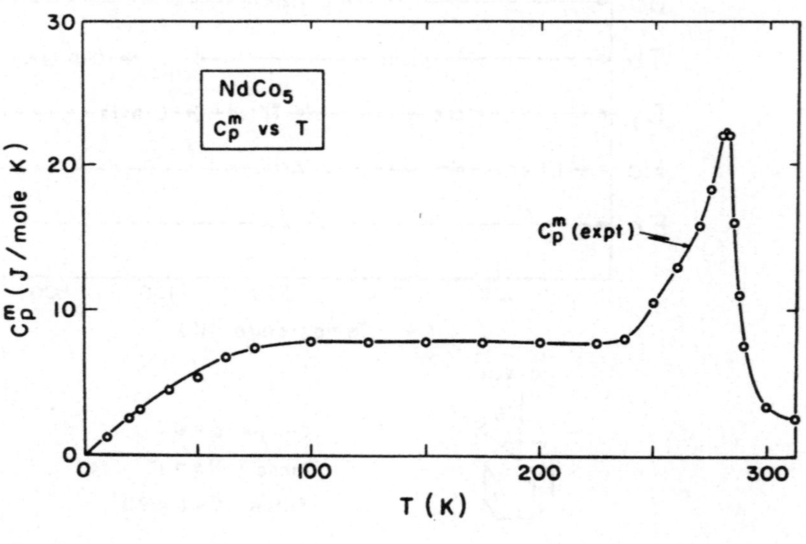

Fig. 18. The excess heat capacity of $NdCo_5$ as a function of temperature.

3. *The R_2Co_{17} Series*

R_2Co_{17} compounds crystallize in two forms – those belonging to the lighter lanthanides crystallize in the Th_2Zn_{17}-type structure and those belonging to the heavier lanthanides, in general, crystallize in the Th_2Ni_{17}-type. Studies of the magnetic properties by *Narasimhan et al.* (*125, 126*), *Miller et al.* (*127, 128*), *Perkins et al.* (*129*) and *Deryagin* and *Kudrevatykh* (*130*) show that the R_2Co_{17} compounds possess high Curie temperatures and reasonably high anisotropies. These data further indicate that the cobalt sublattice in R_2Co_{17} compounds possesses the planar anisotropy. In materials such as Sm_2Co_{17} the rare earth sublattice anisotropy is strongly directed along the c-axis. *Greedan* and *Rao* (*48*) performed theoretical calculations and evaluated the directional preference and stabilization energy of the rare earth ion as a function of temperature. The model adopted by them is the same as that described in the earlier section on RCo_5 compounds. These results are shown in Tables 13–15. There appears to be a general agreement between their calculations and experimental observation. It may be noted that for compounds which crystallize in Th_2Ni_{17}-type structure, the rare earth ion is located in two inequivalent positions (see Fig. 19). The quantity B_2^0 for these types of ions is different. However, the stabilization energy of the site (I) rare earth ions is greater than that of the site (II) ions and hence dominates in determining the anisotropy. In this respect the calculational features resemble those

Table 13. Crystal Field Parameters and Directional Preference for R_2Co_{17} Compounds-Th_2Zn_{17} Structure (*48, 125–130*).

R	$g\mu_B H_{mol}$ (K)	$A_2^0\langle r^2\rangle$ (K)	B_2^0 (K)	Rare Earth Sublattice Preferred Direction (Theory)	(Expt.)
Pr	84	– 325	+ 1.3	plane	(plane)
Nd	115	– 300	+ 0.40	plane	(plane)
Sm	–	– 280	– 2.40	axis	axis
Tb	210	– 225	+ 0.43	plane	(plane)

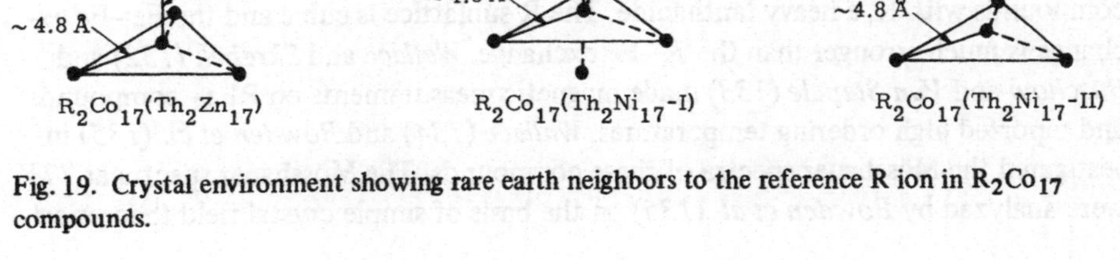

Fig. 19. Crystal environment showing rare earth neighbors to the reference R ion in R_2Co_{17} compounds.

Table 14. Crystal Field Parameters and Directional Preference for R_2Co_{17} Compounds-Th_2Ni_{17} Structure *(48, 125–130)*.

		Type (I) Sites			
R	$g\,\mu_B\,H_{mol}$ (K)	$A_2^0\langle r^2\rangle$ (K)	B_2^0 (K)	Rare Earth Sublattice Preferred Direction	
				(Theory)	(Expt.)
Ho	104	– 1730	+ 0.70	plane	(plane)
Er	84	– 1690	– 0.90	axis	axis
Tm	70	– 1575	– 3.20	axis	axis
Yb	60	– 1520	– 9.5	axis	axis

Table 15.

		Type (II) Sites			
R	$g\,\mu_B\,H_{mol}$ (K)	$A_2^0\langle r^2\rangle$ (K)	B_2^0 (K)	Rare Earth Sublattice Preferred Direction	
				(Theory)	(Expt.)
Ho	104	+ 1315	– 0.52	axis	(plane)
Er	84	+ 1250	+ 0.65	plane	axis
Tm	70	+ 1200	+ 2.4	plane	axis
Yb	60	+ 1160	+ 7.2	plane	axis

discussed in connection with RNi_3 compounds (section IV B 2). Point charge model calculations for B_2^0 in R_2Co_{17} predict the experimental observations reasonably well.

From the foregoing discussion of the R_2Co_{17} and RCo_5 compounds it is clear that the direction of magnetization can immediately be used to establish the sign of B_2^0.

D. *The Rare Earth-Iron Compounds*

The only family of the iron compounds in which considerable work pertaining to the crystal field interaction has been done seems to be those crystallizing in cubic Laves phase structure with the composition RFe_2 *(131)*. With the exception of $CeFe_2$, the lighter lanthanides do not form in this phase. Therefore our discussion is restricted to compounds with R, a heavy lanthanide. The R sublattice is cubic and the Fe–Fe exchange is much stronger than the R–Fe exchange. *Wallace* and *Skrabek* (*132*) and *Buschow* and *Van Stapele* (*133*) made magnetic measurements on RFe_2 compounds and reported high ordering temperatures. *Wallace* (*134*) and *Bowden et al.* (*135*) investigated the Mössbauer spectra of these compounds. The Mössbauer spectra at 77 K were analyzed by *Bowden et al.* (*135*) on the basis of simple crystal field theoretical

Table 16. Magnetic moment directional preference in R Fe_2 compounds. The calculated values are from a consideration of only Fourth Order Parameter (*135–137*).

R	B_4	B_6	Direction of Magnetization in RFe_2	
			Theory Dir.	Expt. Dir. at 77 K
Ce^{4+}	0	0	⟨111⟩	complex
Sm^{3+}	+	0	⟨110⟩	⟨110⟩
Tb^{3+}	+	+	⟨111⟩	⟨111⟩
Dy^{3+}	–	–	⟨100⟩	⟨100⟩
Ho^{3+}	–	+	⟨100⟩	⟨100⟩
Er^{3+}	+	–	⟨111⟩	⟨111⟩
Tm^{3+}	+	+	⟨111⟩	⟨111⟩
Y	0	0	⟨111⟩	⟨111⟩

model. It was found that the direction of magnetization is determined by the sign of the parameter B_4 in Eq. (6). These results are summarized in Table 16.

Results of magnetization on single crystals of some RFe_2 compounds by *Clark et al.* (*136*) and by *Bargouth* and *Will* (*137*) showed that they have very large magnetocrystalline anisotropies. From a comparison of the anisotropy of $TbFe_2$, $DyFe_2$, $HoFe_2$, $ErFe_2$ and $TmFe_2$ with that of YFe_2 it was reported that the huge anisotropy exhibited by the former compounds is a consequence of the CFI. The temperature dependence of microscopic anisotropic constants K_ℓ^0 was examined in detail by *Callen* and *Callen* (*138*). They showed that according to the single ion crystal field model

$$K_\ell^0(T) = K_\ell^0(0)\,[M(T)/M(0)]^{\ell(\ell+1)/2} \qquad (46)$$

where M(T) is the magnetization at temperature T and $K_\ell^0(0)$ is the anisotropy at absolute zero. The temperature dependence of the anisotropy of RFe_2 compounds follow Eq. (46). Results for a variety of compounds were reviewed by *Darby* and *Isaac* (*139*) and by *Clark* (*140*).

Callen and *Callen* (*138*) have also formulated the theory of magnetostriction arising from single ion crystal field effects. This theory accounted satisfactorily for the temperature dependence of (the very large) magnetostriction in RFe_2 compounds. *Clark* (*140*) has successfully utilized the theoretical ideas involving the effect of crystal field interaction to synthesize materials with the most beneficial magnetostrictive and anisotropic properties.

Atzmony et al. (*141, 142*) studied the Mössbauer spectra of $R_x R'_{1-x} Fe_2$ (where R = Ho, Dy and R′ = Tb, Er or Tm) at various temperatures between 4.2 and 300 K. As an illustration, we have reproduced their results on $Ho_x Tb_{1-x} Fe_2$ in Fig. 20.

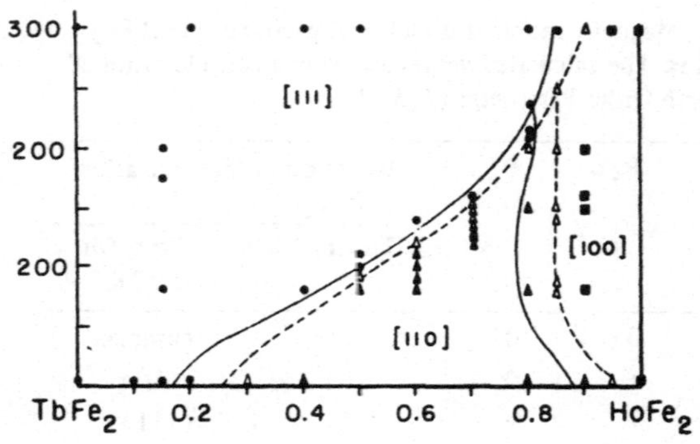

Fig. 20. Spin-orientation diagram in the $Ho_x Tb_{1-x} Fe_2$ system. Filled circles, filled triangles and filled squares correspond to experimentally determined Mössbauer spectra characteristic of the ⟨111⟩, ⟨110⟩ and ⟨100⟩ axes of magnetization, respectively. Open triangles correspond to intermediate types of spectra. The broken lines are the experimentally, and the solid lines theoretically, determined boundaries of regions with different directions of magnetization.

The easy direction of magnetization was determined from the Mössbauer spectra. Their experimental data concerning easy direction of magnetization are described in terms of (x, T) spin-orientation diagrams. Theoretical spin orientation diagrams were calculated assuming that the magnetocrystalline anisotropy was due to the anisotropy of the interaction between the 4f electrons of the rare earth ions with the crystalline fields. The general features of the experimental results were reproduced by this theory. The contribution of the iron sublattice was also calculated and this improved the agreement between the theory and experiment. Crystal field parameters deduced using a point charge model showed that a charge between + 1 and + 2 on the iron atom best describes the experimental features of spin orientations.

E. The Rare Earth-Hydrogen Compounds

The rare earth metals form hydrides readily and a number of studies have been performed on these materials (*143*). The dihydrides of the rare earths form in CaF_2-type structure. The crystal field problem has been applied to characterize the fundamental role of hydrogen in these compounds (*144–149*). The information accumulated for several dihydrides seems to favor the anionic model, namely, that hydrogen accepts an electron to become negatively charged. We will illustrate the results on PrH_2 and ErH_2.

Inverse susceptibility results of PrH_2 shown in Fig. 21 are compared with the values calculated [Eq. (25)] using the spectrum shown in Fig. 22 for anionic hydrogen and an overall crystal field splitting of 657 K. The excellent agreement between experiment and theory seems to support the anionic model for hydrogen with Γ_5 triplet

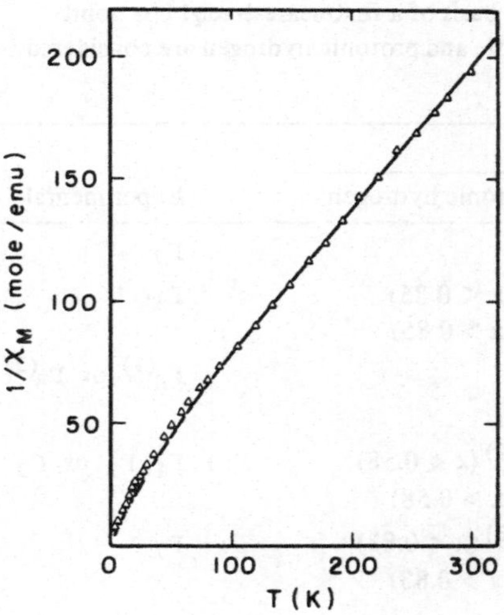

Fig. 21. Comparison of calculated and experimental inverse susceptibility of PrH_2 as a function of temperature. Triangles represent experimental data. The smooth curve is calculated using the level scheme shown in Fig. 22 for anionic hydrogen with Γ_5 as the ground state and CFOAS = 657 K.

as the ground state (*144*). The Schottky heat capacity results reported by *Bieganski* (*145*) also agree with the above results.

Mössbauer experiments of ErH_2 at 4.2 K with external fields up to 35 kOe and in zero field in the temperature range 1.7–77 K were recently reported by *Shenoy et al.* (*146*). They interpreted these results on the basis of an anionic model in which the ground level is a Γ_6 Kramer's doublet (see Fig. 1). Their results agreed reasonably well with the magnetic susceptibility data reported by *Kubota* and *Wallace* (*147*). *Shenoy et al.* (*146*) also calculated the ground level anticipated on the basis of a first nearest-neighbor point charge model for several rare earth dihydrides. These results, for both protonic and anionic models are shown in Table 17 and compared with experimental results. In general, all these results appear to favor an anionic model for hydrogen in RH_2 systems.

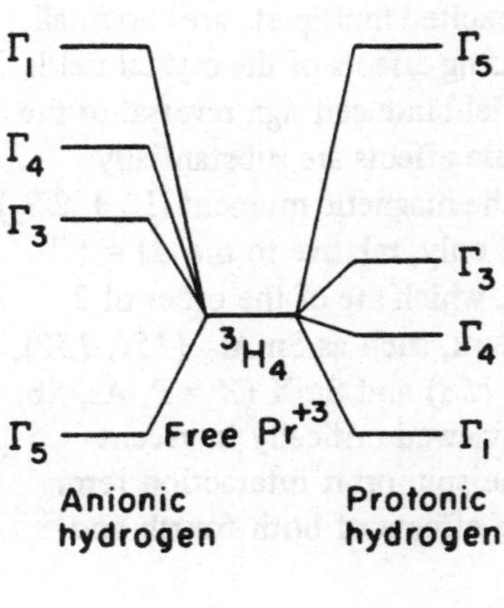

Fig. 22. Splitting pattern of Pr^{3+} ion in PrH_2. calculated for anionic and protonic hydrogen. The energy scale is in arbitrary units and is for a pure fourth order interaction.

Table 17. Crystal field ground state anticipated on the basis of a first-nearest-neighbor point-charge model for the rare-earth dihydrides. Both hydridic and protonic hydrogen are considered, and the known experimental results are shown. (*146*).

	Calculated		
	Hydridic hydrogen	Protonic hydrogen	Experimental
CeH_2	Γ_7	Γ_8	Γ_7
PrH_2	$\Gamma_1 (x < 0.52)$ $\Gamma_5 (x > 0.52)$	$\Gamma_3 (x < 0.85)$ $\Gamma_1 (x > 0.85)$	Γ_5
NdH_2	$\Gamma_8^{(2)}$	Γ_6	$\Gamma_8^{(1)}$ or $\Gamma_8^{(2)}$
SmH_2	Γ_8	Γ_7	
TbH_2	$\Gamma_2 (x < 0.81)$ $\Gamma_3 (x > 0.81)$	$\Gamma_5^{(1)} (x < 0.58)$ $\Gamma_1 (x > 0.58)$	Γ_1, Γ_2 or Γ_3
DyH_2	$\Gamma_7 (x < 0.58)$ $\Gamma_8^{(1)} (x > 0.58)$	$\Gamma_8^{(3)} (x < 0.83)$ $\Gamma_6 (x > 0.83)$	Γ_7
HoH_2	$\Gamma_5^{(2)} (x < -0.54)$ $\Gamma_3^{(2)} (x > -0.54)$	$\Gamma_1 (x < -0.46)$ $\Gamma_5^{(1)} (x > -0.46)$	
ErH_2	$\Gamma_6 (x < -0.46)$ $\Gamma_7 (x > -0.46)$	$\Gamma_8^{(3)}$	Γ_6
TmH_2	$\Gamma_2 (x < 0.81)$ $\Gamma_3 (x > 0.81)$	$\Gamma_5^{(1)} (x < 0.58)$ $\Gamma_1 (x > 0.58)$	

F. The Samarium Compounds

1. Influence of J-Mixing

For a samarium ion in its trivalent state the ground multiplet has $J = 5/2$. Its situation is unlike that in most of the rare earth ions, in that the first excited multiplet is located only about 1400 K above the ground multiplet (*36*). As pointed out by *White* and *Van Vleck* (*150*) admixtures of the ground and excited multiplets are too small to account by themselves for the dramatic and surprising effects of the crystal fields on the magnetic properties of Sm^{3+}, such as crystal field induced sign reversal of the moment relative to the spin at low temperatures. These effects are substantially enlarged by the smallness of the matrix elements of the magnetic moment $(L_z + 2 S_z)$ within the $J = 5/2$, which are of the order of $g_J = 2/7$ only, relative to the $\Delta J = \pm 1$ off-diagonal elements connecting $J = 5/2$ and $J = 7/2$, which are of the order of 2.

The nature of the CFI in compounds of samarium, such as $SmAl_2$ (*151, 152*), $SmPt_2$ (*153, 154*), $SmAl_3$ (*155*), $SmSn_3$ (*153, 156–158*) and SmX (X = P, As, Sb, Bi) (*159, 160*), have been examined in detail and reviewed critically in recent years (2, 3). In general, it has been observed that the spin-orbit interaction term plays a significant role, and in cubic compounds the effects of both fourth and sixth order terms are appreciable. In this section we will illustrate the importance of crystal field interaction in two materials, viz. $SmFe_2$ and $SmCo_5$.

2. *Spin Reorientation in $SmFe_2$*

Employing Mössbauer technique, *Atzmony et al.* (*161*) and *Van Diepen et al.* (*162*) have reported that at temperatures below 80 K the easy direction of magnetization of the samarium sublattice in $SmFe_2$ is directed along ⟨110⟩ direction while at room temperature it is along the ⟨111⟩ direction. The occurrence of spin reorientation in this material as a function of temperature was carefully studied by *Atzmony et al.* (*161*) and *Van Diepen et al.* (*162*). The latter authors have performed detailed calculation on the basis of single ion crystal field model employing a Hamiltonian of the form

$$\mathcal{H} = \lambda \vec{L} \cdot \vec{S} + \mathcal{H}_{CF} + \mathcal{H}_{exch} \tag{47}$$

where the spin-orbit coupling parameter is taken as $\lambda/k = 410$ K. The exchange field was calculated from a knowledge of the magnetic properties of the RFe_2 compounds. Free energy calculations were made for the three possible directions in a cube using Eq. (6), (8) and (9) for various combinations of the fourth and sixth order crystal field parameters (*163*). Fig. 23 shows the regions of easy directions of magnetization as a function of crystal field parameters for four temperatures and an exchange field of $\mu_B H_{ex}/k = 130$ K. It was established that to account for the experimentally observed easy directions of magnetization, $A_4 \langle r^4 \rangle$ and $A_6 \langle r^6 \rangle$ are approximately the same in magnitude, with $A_4 \langle r^4 \rangle$ positive and $A_6 \langle r^6 \rangle$ negative. It was also shown that the sixth order crystal field parameter plays a significant role in determining the behavior of $SmFe_2$. Inclusion of a small amount of anisotropy due to the iron sublattice has negligible effect on the easy axis of magnetization at low temperatures.

3. *Magnetocrystalline Anisotropy in $SmCo_5$*

It is now well known that $SmCo_5$ is useful as a powerful permanent magnet. This property is mainly attributed to its high uniaxial magnetic anisotropy. In an earlier section we have pointed out the systematic dependence of the direction of magnetization with the sign of the crystal field parameter (B_2^0). *Recently Sankar et al.* (*164*)[4]) carried out a detailed analysis of the origin of the magnetocrystalline anisotropy in $SmCo_5$ and its temperature dependence based on the single ion model. Effects of the

[4]) These calculations were performed independently by *Buschow et al.* (*165*).

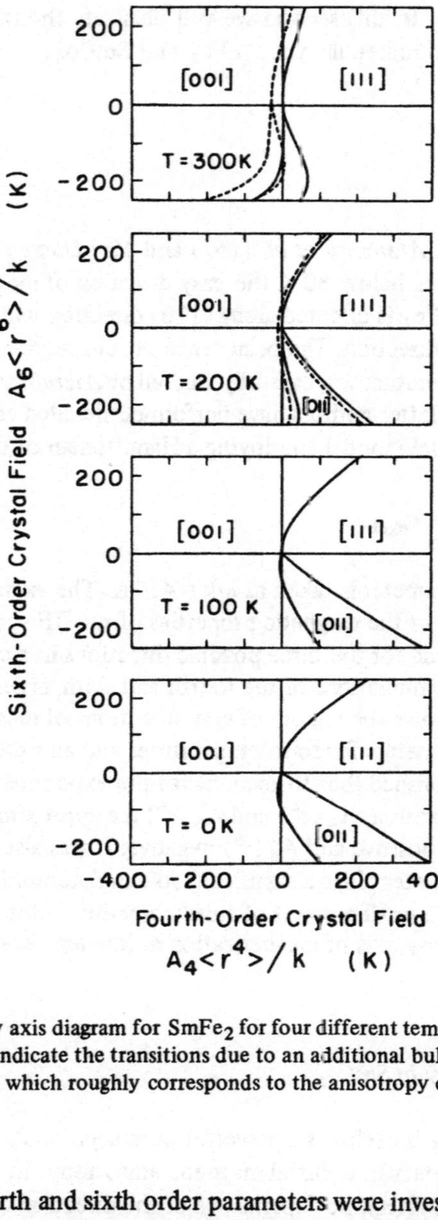

Fig. 23. Easy axis diagram for $SmFe_2$ for four different temperatures and $\mu_B H_{ex}/k = 130$ K. The dashed lines indicate the transitions due to an additional bulk anisotropy of $K_1/k = -2$ K per formula unit, which roughly corresponds to the anisotropy of the iron sublattice.

second, fourth and sixth order parameters were investigated. It was found that the following parameters best describe the experimental magnetic anisotropy of the samarium sublattice:

$$A_2^0 \langle r^2 \rangle = -420\,\mathrm{K},\ A_4^0 \langle r^4 \rangle = -25\,\mathrm{K},\ A_6^0 \langle r^6 \rangle = 1\,\mathrm{K},\ A_6^6 \langle r^6 \rangle = 6\,\mathrm{K}$$

and

$$\mu_B H_{ex}/k = 240\,\mathrm{K}\,.$$

The magnetocrystalline anisotropy calculated for a variety of combinations of crystal field parameters is shown in Fig. 24. It is noted that the second order crystal field term and the exchange term are the most dominant parameters.

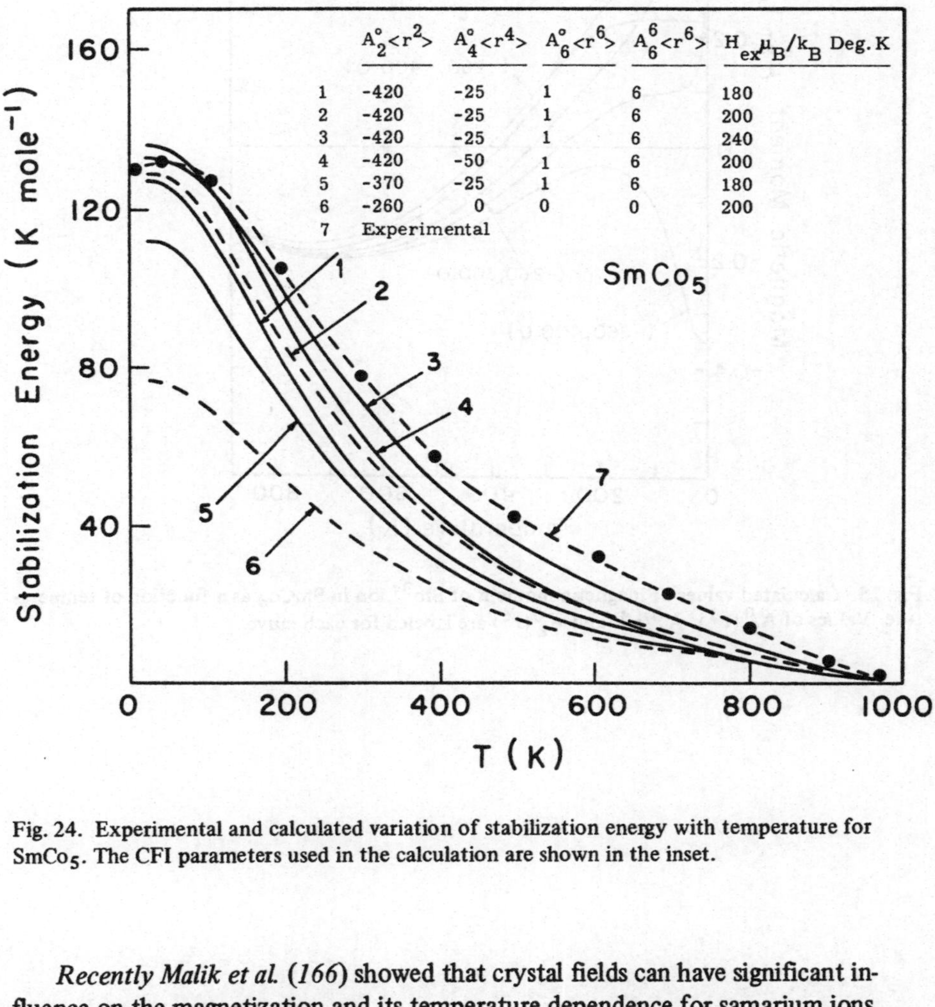

Fig. 24. Experimental and calculated variation of stabilization energy with temperature for $SmCo_5$. The CFI parameters used in the calculation are shown in the inset.

Recently Malik et al. (*166*) showed that crystal fields can have significant influence on the magnetization and its temperature dependence for samarium ions. Their calculations predict a crossover or a sign change of magnetization at a temperature between 400 and 450 K. This means that the coupling between μ_{Sm} and μ_{Co} in $SmCo_5$, which is presumably parallel at low temperatures, becomes antiparallel above the crossover temperature. Their results for a few parameters are shown in Fig. 25. It is seen that for some combinations of crystal fields the crossover temperature can be completely suppressed. Accurate neutron scattering experiments on $SmCo_5$ are being performed by *Schweizer* (*167*) to examine this behavior.

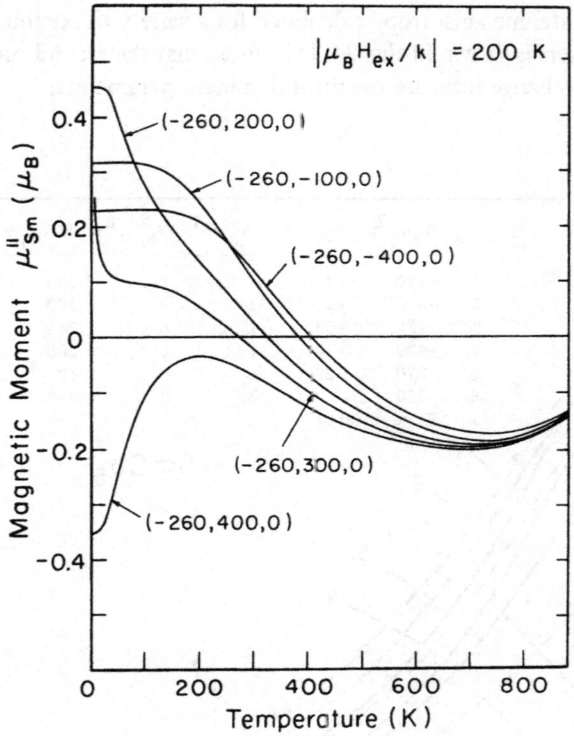

Fig. 25. Calculated values of magnetic moment of Sm^{3+} ion in $SmCo_5$ as a function of temperature. Values of $A_2^0\langle r^2\rangle$, $A_4^0\langle r^4\rangle$ and $A_6^0\langle r^6\rangle$ are labeled for each curve.

V. Concluding Remarks

The crystal field interaction (CFI) modifies, in some cases dramatically, the magnetic properties, thermodynamic behavior and transport properties of rare earth intermetallics. As a consequence of studies carried out during the preceding decade it is now possible to relate these properties to the quantum mechanics of the assemblage of rare earth ions. This is done using the formalism first introduced by *Stevens* in 1952, the so-called equivalent operator method. With this formalism one can derive the crystal field intensity parameters from observations of the aberrations in the magnetic susceptibility (or saturation magnetization) brought on by the CFI, from the heat capacity anomaly associated with excitation of ions within the crystal field spectrum or from the temperature dependence of the spin-disorder resistivity. In a few cases, e.g., $PrAl_3$, the three approaches have been followed, permitting tests of the consistency of the independently derived parameters.

A number of the rare earth intermetallics are very strongly magnetically anisotropic. This leads to useful new permanent magnet materials and substances with very interesting magnetostrictive characteristics. One of the great triumphs of crystal field theory as applied to metallic rare earth systems has been the explanation, provided almost simultaneously by the group working in the Netherlands and at the University of Pittsburgh, of the powerful anisotropy of $SmCo_5$. This is the special feature of this compound which makes it of such considerable significance as a permanent magnet material. This case and a number of other interesting cases have been discussed in this review. It has been shown that observations in regard to changes in anisotropy can be used with the equivalent operator formalism to establish details of the CFI.

The point charge model, to which numerous references have been made in the literature and in this review, emerges as partially successful in accounting for the CFI in that it often leads to the right sign for the crystal field intensity parameters. However, it seldom gives the right magnitude.

The CFI sometimes leads to singlet ground state systems, e.g., $PrNi_2$ or $PrNi_5$ and Van Vleck paramagnetism at low temperatures. These singlet state systems are of intrinsic interest as overt manifestations of the CFI. This phenomenon has been made use of in achieving very low temperatures by bringing about nuclear adiabatic demagnetization.

An interesting consequence of studies of the CFI in rare earth hydrides is that it serves to provide information about the oft-disputed electronic state of hydrogen, whether it is protonic or anionic. For those systems which have been studied by this approach the unambiguous conclusion is that hydrogen is anionic.

Acknowledgement. The present work was assisted by a grant from the U.S. Army Research Office.

References

1. *Cooper, B.R.:* Magnetic Properties of Rare Earth Metals, p. 17 (*Elliott, R.J.*, ed.). New York: Plenum Press 1972 and *Rhyne, J.J.*, ibid., p. 129.
2. *Malik, S.K., Vijayaraghavan, R., Garg, S.K., Ripmeester, R.J.:* Pure and Applied Chemistry **40**, 223 (1974).
3. *de Wijn, H.W., Van Diepen, A.M., Buschow, K.H.J.:* Phys. Stat. Sol. **b76**, 11 (1976).
4. *Lüthi, B.:* Paper presented at the Conference on Magnetism and Magnetic Materials, Pittsburgh (1976).
5. *Wallace, W.E.:* Proc. of the Conf. on Crystal Field Effects in Metals and Alloys, Montreal, p. 1 (1974).
6. *Cochrane, R.W., Harris, R., Plischke, M., Zobin, D., Zuckermann, M.J.:* ibid., p. 79 (1974).
7. *Millhouse, A.H., Furrer, A.:* Proc. of the Annual Conf. on Magnetism and Magnetic Magerials, Philadelphia, p. 257 (1975).
8. *Ruderman, M.A., Kittel, C.:* Phys. Rev. **96**, 99 (1954).
9. *Kasuya, T.:* Progr. Theor. Phys. (Kyoto) **16**, 45 (1956).
10. *Yosida, K.:* Phys. Rev. **106**, 893 (1957).
11. *de Gennes, P.G.:* J. Phys. Radium **23**, 510 (1962).
12. *Kasuya, T.:* Magnetism, IIB, p. 215 (*Rado, G.T., Suhl, H.*, ed.). New York: Academic Press 1966.
13. *Elliott, R.J.:* Magnetic Properties of Rare Earth Metals, p. 1 (*Elliott, R.J.*, ed.). New York: Plenum Press 1972.
14. *Rocher, Y.A.:* Advan. Phys. **11**, 233 (1962).
15. *Kittel, C.:* Solid State Physics, p. 1 (*Seitz, F., Turnbull, D.*, ed.), Vol. 22. New York: Academic Press 1968.
16. *Bethe, H.A.:* Ann. Physik **5**, 133 (1929).
17. *Bleaney, B., Stevens, K.W.H.:* Rep. Progr. Phys. **16**, 108 (1953).
18. *Stevens, K.W.H.:* Proc. Phys. Soc. (Lond.) **A218**, 553 (1953).
19. *Hutchings, M.T.:* Solid State Physics, p. 227 (*Seitz, F., Turnbull, D.*, ed.), Vol. 16. New York: Academic Press 1966.
20. *Elliot, R.J., Stevens, K.W.H.:* Proc. Roy. Soc. (Lond.) **A218**, 553 (1953) and *Freeman, A.J., Watson, R.E.:* Phys. Rev. **127**, 2058 (1962).
21. *Burns, G.:* J. Chem. Phys. **42**, 377 (1965).
22. *Das, K.C., Ray, D.K.:* Solid State Comm. **8**, 2025 (1970).
23. *Jørgensen, C.K., Pappalardo, R., Smidtke, H.H.:* J. Chem. Phys. **39**, 1422 (1963).
24. *Lea, K.R., Leask, M.J.M., Wolf, W.P.:* J. Phys. Chem. Solids **23**, 1381 (1962).
25. *Segal, E., Wallace, W.E.:* J. Solid State Chem. **2**, 347 (1970).
26. *Racah, G.:* Phys. Rev. **62**, 438 (1942); **63**, 367 (1943).
27. *Weber, M.J., Bierig, R.W.:* ibid., **134**, A1502 (1964).
28. *Rotenberg, M., Bivins, R., Metropolis, N., Wooten, J.K.:* The 3–j and 6–j Symbols. Cambridge, Mass.: MIT Press 1959.
29. *Nielson, C.W., Koster, G.F.:* Spectroscopic Coefficients for p^n, d^n and f^n Configurations. Cambridge, Mass.: MIT Press 1964.
30. *Schlesinger, M., Nara, H.:* J. Phys. Chem. Solids **34**, 1827 (1973).
31. *Furrer, A., Tellenbach, U.:* Helv. Phys. Acta. **48**, 451 (1975).
32. *Elliott, J.P., Judd, B.R., Runciman, W.A.:* Proc. Roy. Soc. (Lond.) **A240**, 509 (1957).
33. *Schumacher, D.P., Hollingsworth, C.A.:* J. Phys. Chem. Solids **27**, 749 (1966).
34. *Ebina, Y., Tsuya, N.:* Sci. Rep. RITU **B12**, 166 (1960); **B13**, 1 (1961) and **B13**, 25 (1961).
35. *Penney, W.G., Schlapp, R.:* Phys. Rev. **41**, 194 (1932).
36. *Van Vleck, J.H.:* The Theory of Electric and Magnetic Susceptibilities. Oxford University Press 1932.
37. *Segal, E., Wallace, W.E.:* J. Solid State Chem. **6**, 99 (1973); **11**, 203 (1974) and **13**, 201 (1975).

38. *Schottky, W.:* Physik. Z., **23**, 448 (1922) and *Jackson, L.C.;* Low Temperature Physics. London: Methuen and Co. Ltd. 1962.
39. *Birgeneau, R.J., Bucher, E., Passell, L., Price, D.L., Turberfield, K.C.:* J. Appl. Phys. **41**, 900 (1970).
40. *Furrer, A.:* Proc. 11. Rare Earth Res. Conf. (Traverse City, Michigan), p. 308 (1974).
41. *Kasuya, T.:* Progr. Theor. Phys. (Kyoto) **22**, 227 (1959).
42. *de Gennes, P.G., Friedel, J.:* J. Phys. Chem. Solids **4**, 71 (1958).
43. *Van Peski-Tinbergen, T., Dekker, A.J.:* Physica **29**, 917 (1963).
44. *Dekker, A.J.:* J. Appl. Phys. **36**, 906 (1965).
45. *Rao, V.U.S., Wallace, W.E.:* Phys. Rev. **B2**, 4613 (1970).
46. *Maranzana, F.E.:* Phys. Rev. Lett. **25**, 239 (1970).
47. *Cornut, B., Coqblin, B.:* Phys. Rev. **B5**, 4541 (1972).
48. *Greedan, J.E., Rao, V.U.S.:* J. Solid State Chem. **6**, 387 (1973); **8**, 368 (1973).
49. *Wernick, J.H., Geller, S.:* Trans. AIME **218**, 866 (1960).
50. *Williams, H.J., Wernick, J.H., Nesbitt, E.A., Sherwood, R.C.:* J. Phys. Soc. Japan, Suppl. B1, **17**, 91 (1962).
51. *White, J.A., Williams, H.J., Wernick, J.H., Sherwood, R.C.:* Phys. Rev. **131**, 1039 (1963).
52. *Mader, K.H., Wallace, W.E.:* J. Chem. Phys. **49**, 1521 (1968).
53. *Nereson, N., Olsen, C., Arnold, G.:* J. Appl. Phys. **37**, 4575 (1966); **39**, 4605 (1968); *Olsen, C.E., Arnold, G., Nereson, N.*, ibid., **38**, 1395 (1967).
54. *Inoue, T., Sankar, S.G., Craig, R.S., Wallace, W.E., Gschneidner, K.A.:* J. Phys. Chem. Solids **38**, 487 (1977).
55. *Inoue, T.:* Ph. D. Thesis, University of Pittsburgh (1975).
56. *Purwins, H.G., Walker, E., Barbara, B., Rossignol, M.F., Bak, P.:* J. Phys. **C7**, 3573 (1974).
57. *Purwins, H.G., Walker, E., Barbara, B., Rossignol, M.F., Furrer, A.:* J. Phys. **C9**, 1025 (1976).
58. *Heer, H., Furrer, A., Walker, E., Treyvaud, A., Purwins, H.G., Kjems, J.:* J. Phys. **C7**, 1207 (1974).
59. *Wiedemann, W., Zinn, W.:* Phys. Lett. **24 A**, 506 (1967).
60. *Devine, R.A.B., Zingg, W., Moret, J.M., Shaltiel, D.:* Solid State Comm. **12**, 515 (1973).
61. *Sankar, S.G., Malik, S.K., Rao, V.U.S.:* J. Solid State Chem. **18**, 303 (1976).
62. *Barbara, B., Rossignol, M.F., Boucherle, J.X.:* Phys. Lett. **55 A**, 321 (1975).
63. *Hill, T.W.:* M.S. Thesis, University of Pittsburgh (1972).
64. *Barbara, B., Boucherle, J.X., Rossignol, M.F., Schweizer, J.:* Paper presented at the "Conference on Neutron Scattering", Gatlinburg (1976).
65. *Wallace, W.E., Craig, R.S., Thompson, A.W., Deenadas, C., Dixon, M., Aoyagi, M., Marzouk, N.:* Les Elements des Terres Rares, Colloque Internationaux du C.N.R.S. (Paris) no. 180, **2**, 427 (1970).
65(a). *Swift, W.M., Wallace, W.E.:* J. Phys. Chem. Solids **29**, 2053 (1968).
66. *Mader, K.H., Segal, E., Wallace, W.E.:* ibid., **30**, 1 (1969).
67. *Cooper, B.R., Wang, Y.L.:* Phys. Rev. **163**, 444 (1967); **172**, 539 (1968) and **185**, 696 (1969).
68. *Barbara, B., Rossignol, M.F., Purwins, H.G., Walker, E.:* Colloque des Champs Intenses, C.N.R.S., p. 51 (1974).
69. *Deenadas, C., Thompson, A.W., Craig, R.S., Wallace, W.E.:* J. Phys. Chem. Solids **32**, 1853 (1971).
70. *Houmann, J.G., Bak, P., Purwins, H.G., Walker, E.:* J. Phys. **C7**, 2691 (1974).
71. *Purwins, H.G., Houmann, J.G., Bak, P., Walker, E.:* Phys. Rev. Lett. **31**, 1585 (1973); *Barbara, B., Rossignol, M.F., Purwins, H.G., Walker, E.:* Phys. Stat. Sol. **22a**, 553 (1974).
72. *Van Vucht, J.H.N., Buschow, K.H.J.:* J. Less Common Metals **10**, 98 (1965).
73. *Buschow, K.H.J., Van Vucht, J.H.N.:* Z. Metallk. **57**, 162 (1966).
74. *Mahoney, J.V., Wallace, W.E., Craig, R.S.:* J. Appl. Phys. **45**, 2733 (1974).
75. *Mahoney, J.V.:* Ph. D. Thesis, University of Pittsburgh (1972).

76. *Mahoney, J.V., Rao, V.U.S., Wallace, W.E., Craig, R.S.:* Phys. Rev. **B9**, 154 (1974).
77. *Rao, V.U.S., Suski, W., Craig, R.S., Wallace, W.E.:* Proc. Rare Earth Res. Conf., Reno, p. 355 (1970).
78. *Van Daal, H.J., Buschow, K.H.J.:* Phys. Stat. Sol. **3**, 853 (1970).
79. *Mahoney, J.V., Wallace, W.E., Craig, R.S., Sankar, S.G.:* Inorg. Chem. **14**, 2918 (1975).
80. *Buschow, K.H.J.:* Z. Phys. Chem. Neue Folge **59**, 21 (1968).
81. *Bucher, E., Maita, J.P., Hull, G.W., Sierro, J., Chu, C.W., Lüthi, B.:* Proc. 1st Conf. on Cryst. Elect. Field Effects in Metals and Alloys, p. 221 (1974).
82. *Mader, K.H., Swift, W.M.:* J. Phys. Chem. Solids **29**, 1759 (1968).
83. *de Gennaro, S., Borchi, E.:* Phys. Rev. Lett. **30**, 377 (1973).
84. *Edelstein, A.S., Tranchita, C.J., McMasters, O.D., Gschneidner, K.A.:* Solid State Comm. **15**, 81 (1974).
85. *Skrabek, E.A., Wallace, W.E.:* J. Appl. Phys. *34*, 1356 (1963).
86. *Bleaney, B.:* Proc. Roy. Soc. (Lond.) **276 A**, 19 (1963).
87. *Wallace, W.E., Mader, K.H.:* Inorg. Chem. **7**, 1627 (1968).
88. *Nowik, I., Dunlap, B.D.:* J. Phys. Chem. Solids **34**, 465 (1973).
89. *Cromer, T., Olsen, O.E.:* Acta Cryst. **12**, 689 (1959).
90. *Paccard, D., Pauthenet, R.:* Compt. Rend. **26 A**, 1056 (1967).
91. *Rossat-Mignod, J., Yakinthos, J.:* Phys. Stat. Sol. **b47**, 239 (1971).
92. *Yakinthos, J., Paccard, D.:* Solid State Comm. **10**, 989 (1972).
93. *Yakinthos, J., Rossat-Mignod, J., Belakhovsky, M.:* Phys. Stat. Sol. **(b)47**, 247 (1971).
94. *Wernick, J.H., Geller, S.:* Acta Cryst. **12**, 662 (1959).
95. *Nesbitt, E.A., Williams, H.J., Wernick, J.H., Sherwood, R.C.:* J. Appl. Phys. **33**, 1674 (1962).
96. *Wallace, W.E., Aoyagi, M.:* Monatshefte für Chem. **102**, 5 (1971).
97. *Craig, R.S., Sankar, S.G. Marzouk, N., Rao, V.U.S., Wallace, W.E., Segal, E.:* J. Phys. Chem. Solids **33**, 2267 (1972).
98. *Marzouk, N., Craig, R.S., Wallace, W.E.:* ibid., **34**, 15 (1973).
99. *Sankar, S.G., Keller, D.A., Craig, R.S., Wallace, W.E., Rao, V.U.S.:* J. Solid State Chem. **9**, 78 (1974).
100. *Andres, K., Schmidt, P.H., Darack, S.:* AIP Proc., No. 24, 238 (1974).
101. *Bleaney, B.:* Proc. Phys. Soc. (Lond.) **82**, 469 (1963).
102. *Ross, J.W., Crangle, J.:* Phys. Rev. **133**, A509 (1964).
103. *Wallace, W.E., Skrabek, E.A.:* Rare Earth Research, p. 431 (*Vorres, K.S.*, editor). New York: Gordon and Breach 1964.
104. *Farrell, J., Wallace, W.E.:* Inorg. Chem. **5**, 105 (1966).
105. *Crangle, J., Ross, J.W.:* Proc. Int. Conf. Magnet., Nottingham, p. 240, London: Inst. of Phys. and Phys. Soc. (1964).
106. *Lemaire, R.:* Cobalt **32**, 132 (1966).
107. *Wallace, W.E.:* Rare Earth Intermetallics. New York: Academic Press 1972.
108. *Schweizer, J.:* Phys. Lett. **24A**, 739 (1967).
109. *Moon, R.M., Koehler, W.C., Farrell, J.:* J. Appl. Phys. **36**, 978 (1965).
110. *da Silva, J.M.M., McDermott, J.M., Hill, R.W.:* J. Phys. **C5**, 1573 (1972).
111. *Thompson, A.W.:* Ph. D. Thesis, University of Pittsburgh (1971).
112. *Dublon, G., Kroupp, M., Dariel, M.P., Atzmony, U.:* Phys. Stat. Sol. **b76**, 669 (1976).
113. *Gignoux, D., Givord, F., Lemaire, R.:* Paper No. L4, Second International Conf. on Crystal Electric Field Effects in Metals and Alloys, Zürich (1976).
114. *Gignoux, D., Givord, F., Lemaire, R.:* Phys. Rev. **B12**, 3878 (1975).
115. *Deportes, J., Gignoux, D., Givord, F.:* Phys. Stat. Sol. **b64**, 29 (1974).
116. *Gignoux, D., Givord, D., Givord, F., Koehler, W.C., Moon, R.M.:* Phys. Rev. **B14**, 162 (1976).
117. *Hoffer, G., Strnat, K.J.:* J. Appl. Phys. **38**, 1377 (1967).
118. *Hoffer, G., Strnat, K.:* IEE Trans. **MAG-2**, 487 (1969).
119. *Tatsumoto, E., Okamoto, T., Fujii, H., Inoue, C.:* J. de Phys. **32**, Cl-550 (1971).
120. *Benz, M.G., Martin, D.L.:* J. Appl. Phys. **43**, 4733 (1972).

121. *Nesbitt, E.A., Wernick, J.H.:* Rare Earth Permanent Magnets. New York: Academic Press 1973.
122. *Okamoto, T., Fujii, H., Inoue, C., Tatsumoto, E.:* J. Phys. Soc. Japan **34**, 835 (1973).
123. *Klein, H.P., Menth, A.:* A.I.P. Proc., No. 18, 1177 (1973).
124. *Keller, D.A., Sankar, S.G., Craig, R.S., Wallace, W.E.:* ibid., No. 18, 1207 (1973).
125. *Narasimhan, K.S.V.L., Wallace, W.E.:* ibid., No. 18, 1248 (1973).
126. *Narasimhan, K.S.V.L., Wallace, W.E., Hutchens, R.D.:* IEEE Trans. **MAG-10**, 729 (1974).
127. *Miller, A.E., Igarashi, H.:* Proc. Rare Earth Res. Conf. (Traverse City, Michigan), p. 439 (1974).
128. *Miller, A.E., D'Silva, T., Miura, K.:* ibid., p. 461 (1974).
129. *Perkins, R.S., Strässler, S., Menth, A.:* A.I.P. Proc., No. 29, 610 (1975).
130. *Deryagin, A.V., Kudrevatikh, N.V.:* Phys. Stat. Sol. **a30**, K129 (1975).
131. *Van der Goot, A.S., Buschow, K.H.J.:* J. Less Common Metals **21**, 151 (1970).
132. *Wallace, W.E., Skrabek, E.A.:* Rare Earth Research, (*Vorres, K.S.*, editor), Vol. 2, p. 431. New York: Gordon and Breach 1964.
133. *Buschow, K.H.J., Van Stepele, R.P.:* J. Appl. Phys. **41**, 4066 (1970).
134. *Wallace, W.E.:* J. Chem. Phys. **41**, 3857 (1964).
135. *Bowden, G.J., Bunbury, D.St.P., Guimaraes, A.P., Snyder, R.E.:* J. Phys. **C2**, 1376 (1968).
136. *Clark, A.E., Belson, H.S., Tamagawa, N.:* Phys. Lett. **42A**, 160 (1972).
137. *Bargouth, M.O., Will, G.:* J. de Phys. **C1 32**,675 (1971).
138. *Callen, E., Callen, H.B.:* Phys. Rev. **139**, A455 (1965).
139. *Darby, M.I., Isaac, E.D.:* IEEE Trans. **MAG-10**, 259 (1974).
140. *Clark, A.E.:* A.I.P. Proc., No. 18, 1015 (1973).
141. *Atzmony, U., Dariel, M.P., Bauminger, E.R., Lebenbaum, D., Nowik, I., Ofer, S.:* Phys. Rev. Lett. **28**, 244 (1972).
142. *Atzmony, U., Dariel, M.P., Bauminger, E.R., Lebenbaum, D., Nowik, I., Ofer, S.:* Phys. Rev. **B7**, 4220 (1973).
143. *Bos, W.G., Gayer, K.H.:* J. Nucl. Mater. **18**, 1 (1960).
144. *Wallace, W.E.:* Bunsenges, Physik Chem. **76, 832 (1972).**
145. *Bieganski, Z.:* J. Chem. Therm. **5**, 1 (1973).
146. *Shenoy, G.K., Dunlap, B.D., Westlake, D.G., Dwight, A.E.:* Phys. Rev. **B14**, 41 (1976).
147. *Kubota, Y., Wallace, W.E.:* J. Appl. Phys. Suppl. **33**, 1348 (1962).
148. *Wallace, W.E., Mader, K.H.:* J. Chem. Phys. **48**, 84 (1968).
149. *Bieganski, Z., Stalinski, B.:* J. Less Common Metals **49**, 421 (1976).
150. *White, J.A., Van Vleck, J.H.:* Phys. Rev. Lett. **6**, 412 (1961).
151. *de Wijn, H.W., Van Diepen, A.M., Buschow, K.H.J.:* Phys. Rev. **B7**, 524 (1973).
152. *Buschow, K.H.J., Van Diepen, A.M., de Wijn, H.W.:* ibid., **B8**, 5134 (1973).
153. *Malik, S.K.:* Phys. Lett. **31A**, 33 (1970).
154. *Malik, S.K., Vijayaraghavan, R., Sankar, S.G., Rao V.U.S.:* (unpublished).
155. *de Wijn, H.W., Van Diepen, A.M., Buschow, K.H.J.:* Phys. Rev. **161**, 253 (1967).
156. *Rao, V.U.S., Vijayaraghavan, R.:* Phys. Lett. **19**, 168 (1965).
157. *Borsa, F., Barnes, R.G., Reese, R.A.:* Phys. Stat. Sol. **19**, 359 (1967).
158. *Buschow, K.H.J., Van Diepen, A.M., de Wijn, H.W.:* Phys. Lett. **24A**, 536 (1967).
159. *Jones, E.D., Hesse, J.E.:* J. Appl. Phys. **38**, 1159 (1967).
160. *Jones, E.D.:* Phys. Rev. **180**, 455 (1969).
161. *Atzmony, U., Dariel, M.P., Bauminger, E.R., Lebenbaum, D., Nowik, I., Ofer, S.:* Proc. Rare Earth Res. Conf., p. 605 (1973).
162. *Van Diepen, A.M., de Wijn, H.W., Buschow, K.H.J.:* Phys. Rev. **B8**, 1125 (1973).
163. *de Wijn, H.W., Van Diepen, A.M., Buschow, K.H.J.:* Proc. 1st Conf. on Cryst. Elect. Field Effects in Metals and Alloys, p. 401, Montreal (1974).
164. *Sankar, S.G., Rao, V.U.S., Segal, E., Wallace, W.E., Frederick, W.G.D., Garrett, H.J.:* Phys. Rev. **B11**, 435 (1975). Also see Montreal Cont. Proc. p. 454 (1974).
165. *Buschow, K.H.J., Van Diepen, A.M., de Wijn, H.W.:* Sol. State Comm. **15**, 903 (1974).
166. *Malik, S.K., Sankar, S.G., Rao, V.U.S., Wallace, W.E.:* Mag. Lett. **1**, 51 (1977).
167. *Schweizer, J.:* (private communication).

Molecular Orbital Bonding Concepts in Polyatomic Molecules: a Novel Pictorial Approach

D.K. Hoffman, K. Ruedenberg and J.G. Verkade,

Department of Chemistry and Ames Laboratory USERDA, Iowa State University, Ames, Iowa 50011

Table of Contents

In recent years accurate molecular orbital wave functions have become available for increasingly complex molecules and, with the development of sophisticated computational techniques, they are yielding useful predictions of molecular electronic properties (*1, 2*). This being so, it is becoming important for the non-theoretically oriented chemist to appreciate the underlying principles governing molecular orbital formation and thereby to predict those results which follow from general principles such as those derived from group theory (*3*) or fundamental concepts of binding and antibinding, and distinguish them from the quantitative details associated with particular molecules. It appears highly desirable, therefore, that the non-theoretician be able to deduce results of general validity without esoteric mathematics, preferably in a pictorial manner. Such an approach is virtually indispensible if bonding concepts are to be taught to chemistry students early in their careers.

A particular principle of bonding which merits sharper focus is the equivalence of different sets of molecular orbitals. In the MO approximation, there exist various sets of molecular orbitals with varying degrees of localization and delocalization which contain exactly the same information about a given molecule, albeit in differing representations, and which can be used interchangeably (*4*). The ability to pictorially manipulate such equivalent MO sets is an extremely useful tool for clarifying the interpretation of bonding patterns, especially if the delocalized as well as the localized description (or even mixed descriptions) are desired for the same molecule.

To our knowledge, apart from a brief and elementary outline of a new approach developed by the present authors (*5*), no simple systematic didactic method for accomplishing the aforementioned goals has been reported (particularly for polyatomic molecules). While excellent introductory descriptions of bonding concepts exist (*6, 7*), no attempts seem to have been made to find a pictorial substitute for a substantial portion of group theory as applied to molecular orbitals or to elaborate in detail the equivalence of the localized and delocalized bonding views on an elementary level. Our approach has been developed and tested in a freshman chemistry course for majors at Iowa State University for a number of years. In the present paper we give and justify a more elaborate discussion of this pictorial method which leads to delocalized and localized MO's for a wide variety of polyatomic molecules. The key concept is that delocalized and localized MO's can be deduced from an appropriate extension of the characteristics of AO's. More specifically, *the symmetry and directional characteristics of MO's are obtained from the symmetry and directional characteristics of AO's.*

1. Valence and Generator Atomic Orbitals

For the purpose of MO construction the most important feature of an AO is its nodal pattern. The sign of the AO amplitude changes each time a nodal surface is crossed. This is analogous to the sign change in the amplitude of a standing wave in a plucked guitar string. In the case of *canonical* AO's there are three types of nodes: spheres (zeros of the radial function), cones concentric with the *z* axis[1]) (zeros of the θ function) and "vertical" planes containing the *z* axis (zeros of the Φ function) (*8*). These are illustrated in Figure 1 for the 2s, 2p and 3d orbitals. We note two points regarding these nodal surfaces: first, a cone with an apex angle of 180° is identical with the *xy* plane (as in 2 p0 = 2 pz); it is convenient to consider this nodal plane as a conal node. Secondly, the vertical planar nodes are evenly spaced about the *z* axis. Canonical AO's with planar nodes always come in pairs. The orbitals of each pair are of identical shape but are rotated with respect to each other by one-half the angular spacing of the nodal plane(s) (or by 90° in the case of one vertical plane). Thus 2 p*x* = 2 p1′ rotated 90° gives rise to 2 p*y* = 2 p1″ as shown in Fig. 1, and 3 d$x^2 - y^2$ = 3 d2′ rotated by 45° produces 3 d*xy* = 3 d2″ as shown in Fig. 2b[2]).

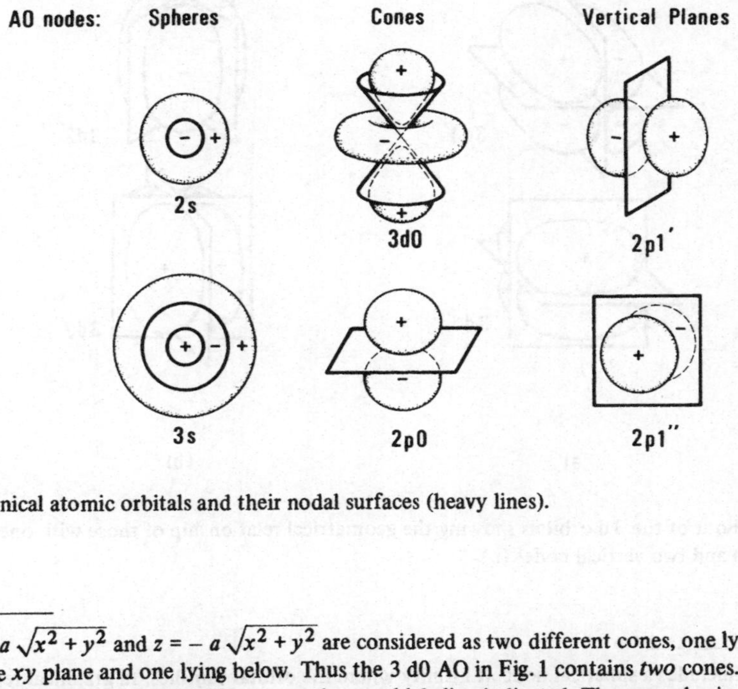

Fig. 1. Canonical atomic orbitals and their nodal surfaces (heavy lines).

---

[1]) Here $z = a\sqrt{x^2 + y^2}$ and $z = -a\sqrt{x^2 + y^2}$ are considered as two different cones, one lying above the *xy* plane and one lying below. Thus the 3 d0 AO in Fig. 1 contains *two* cones.

[2]) Since we deal only with real AO's, we use the novel labeling indicated. The general principle of this notation is that canonical AO's containing cos $m\Phi$ are labeled by m' and those containing sin $m\Phi$ are indicated by m''.

The nodal pattern of any canonical AO is related to its quantum numbers (nlm') or (nlm''). We begin with the general rule that the total number of nodes is n.[3]) Starting with the spherical orbital (n00) = ns, the nonspherical orbitals can be generated as shown for $n = 3$ in Fig. 1 to 3. The 3s = (300) AO has three spherical nodes. Converting one spherical node to a conal node gives us a 3 p0 = 3 pz orbital, while converting it to a vertical planar node leads to the 3 p1′ and 3 p1″ AO's as shown in Fig. 3. Similarly, converting the two spherical nodes to two conal nodes gives us a 3 d0 = 3 dz^2 AO (Fig. 1) and, changing the conal nodes one at a time to vertical planar nodes, generates the remaining four 3 d orbitals (Fig. 2). The radial node at infinity must not be converted to an angular node. Thus by beginning with the total number of nodes for any given quantum number in spherical form and converting them one at a time to conal and then to vertical nodes, all the possible nodal combinations can be realized in a systematic way as illustrated in Table 1. The resulting canonical AO's shown in Figs. 1–4 occupy the spaces between the nodal surfaces and their amplitudes can be visualized in three-dimensional form by sketching spheres, lobes or toruses to accommodate the geometries of these surfaces. As usual *hybrid* AO's are defined as linear combinations of canonical AO's having different l quantum numbers.

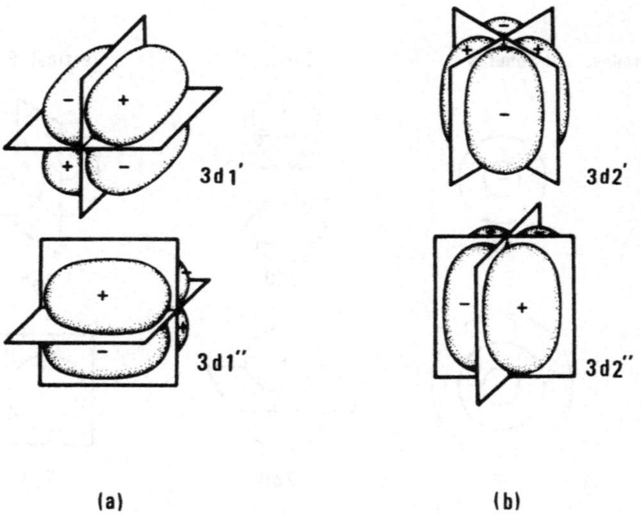

Fig. 2. Four of the 3 d orbitals showing the geometrical relationship of those with one vertical node (a) and two vertical nodes (b).

[3]) This includes a spherical node at infinity where the orbital vanishes. In graphical representations this node is replaced by a limiting *contour* of small orbital density which encompasses some fixed large fraction of the electron density. This contour is no longer necessarily spherical; it depicts the "shape" of the orbital.

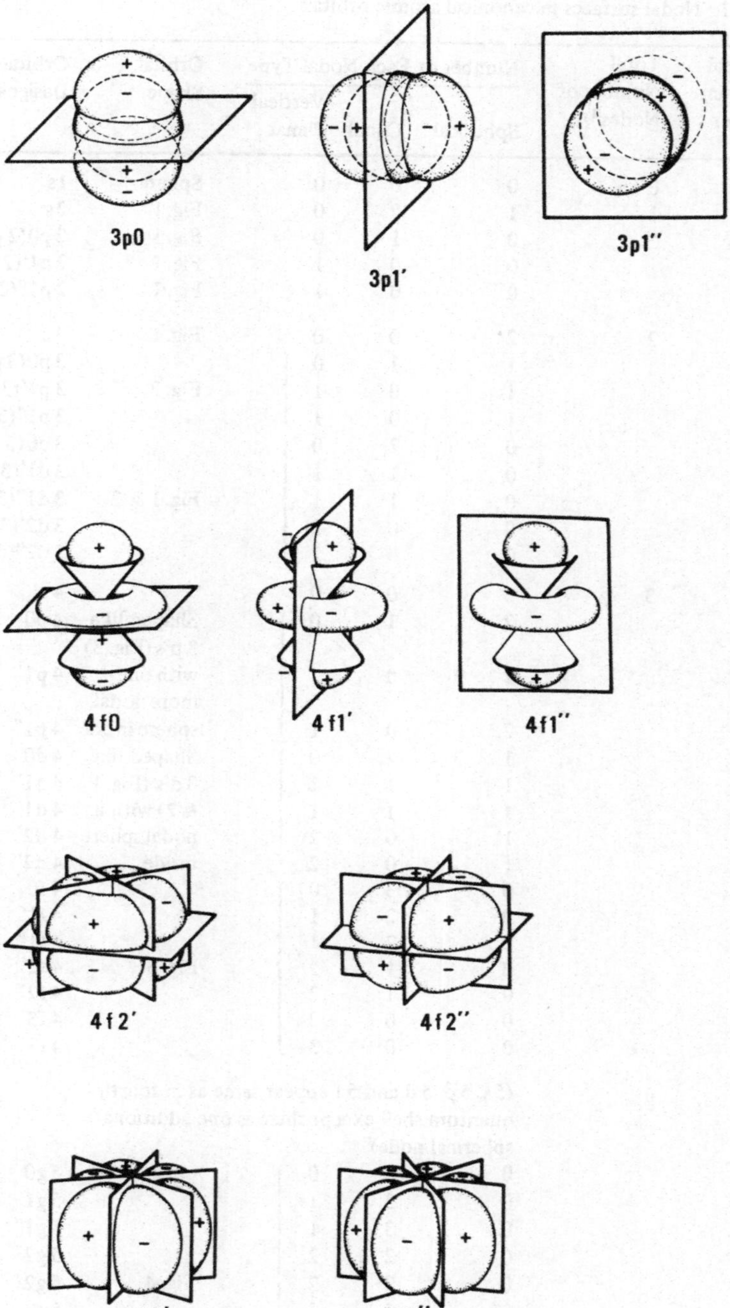

Fig. 3. Representations of the 3 p and 4 f atomic orbitals showing the nodal surfaces (dotted and heavy lines).

Table 1. Nodal surfaces in canonical atomic orbitals

Principal Quantum Number	Total Number of Nodes[a])	Number of Each Nodal Type			Orbital Shape	Orbital Designation
		Spherical	Conal	Vertical Planar		
1	0	0	0	0	Sphere	1s
2	1	1	0	0	Fig. 1	2s
		0	1	0	Fig. 1	2 p0(2 pz)
		0	0	1	Fig. 1	2 p1′(2 px)
		0	0	1	Fig. 1	2 p1″(2 py)
3	2	2	0	0	Fig. 1	3 s
		1	1	0		3 p0(3 pz)
		1	0	1	Fig. 3	3 p1′(3 px)
		1	0	1		3 p1″(3 py)
		0	2	0		3 d0(3 d $3z^2-r^2$)
		0	1	1		3 d1′(3 dxz)
		0	1	1	Fig. 1 & 2	3 d1″(3 dyz)
		0	0	2		3 d2′(3 dx^2-y^2)
		0	0	2		3 d2″(3 dxy)
4	3	3	0	0		4 s
		2	1	0	Shaped like 3 p's (Fig. 3)	4 p0
		2	0	1	with one more nodal	4 p1′
		2	0	1	sphere inside	4 p1″
		1	2	0	Shaped like	4 d0
		1	1	1	3 d's (Fig. 1	4 d1′
		1	1	1	& 2) with a	4 d1″
		1	0	2	nodal sphere	4 d2′
		1	0	2	inside	4 d2″
		0	3	0		4 f0
		0	2	1		4 f1′
		0	2	1		4 f1″
		0	1	2	Fig. 3	4 f2′
		0	1	2		4 f2″
		0	0	3		4 f3′
		0	0	3		4 f3″
5	4	(5 s, 5 p, 5 d and 5 f appear same as in fourth quantum shell except there is one additional spherical node)				
		0	4	0		5 g0
		0	3	1		5 g1′
		0	3	1		5 g1″
		0	2	2		5 g2′
		0	2	2	Fig. 4	5 g2″
		0	1	3		5 g3′
		0	1	3		5 g3″
		0	0	4		5 g4′
		0	0	4		5 g4″

[a]) Not including the node at infinity.

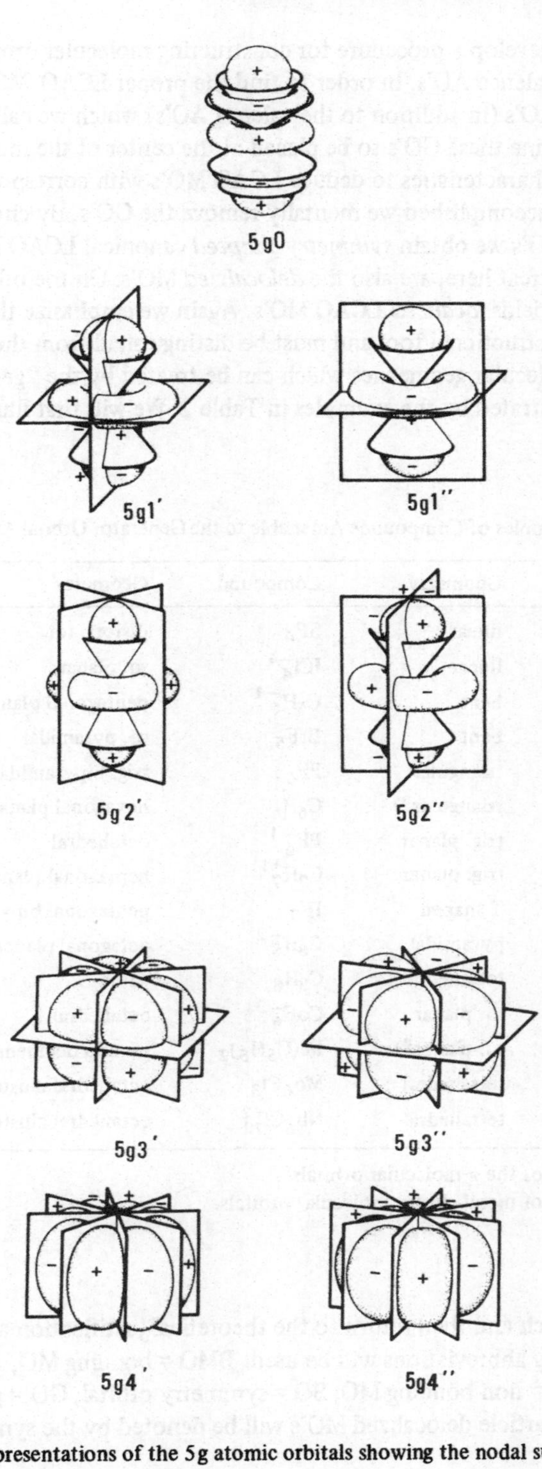

Fig. 4. Representations of the 5g atomic orbitals showing the nodal surfaces (heavy lines).

Our objective is to develop a procedure for constructing molecular orbitals as linear combinations of valence AO's. In order to find the proper LCAO MO's, we introduce a new set of AO's (in addition to the valence AO's) which we call generator orbitals (GO's). We imagine these GO's to be placed at the center of the molecule and we use their symmetry characteristics to deduce LCAO MO's with corresponding properties. After this is accomplished we mentally remove the GO's. By choosing appropriate *canonical* GO's we obtain *symmetry adapted* canonical LCAO MO's which, for the cases we treat here, are also the *delocalized* MO's. On the other hand, the use of *hybrid* GO's yields *localized* LCAO MO's. Again we emphasize that the GO's serve only as a constructional tool and must be distinguished from the valence AO's. The variety of molecular geometries which can be treated by the "generator orbital" approach is illustrated by the examples in Table 2. We will first illustrate the

Table 2. Examples of Compounds Amenable to the Generator Orbital Approach

Compound	Geometry	Compound	Geometry
BeF_2	linear	SF_4	distort. tet.
XeF_2	linear	ICl_4^{-1}	sq. planar
H_2O	bent	$C_5H_5^{-1}$	pentagonal planar[a])
NO_2^{-1}	bent	BrF_5	sq. pyramidal
H_3^+	triangular	PF_5	trig. bipyramidal
$C_3H_3^{+1}$	triangular[a])	C_6H_6	hexagonal planar
CH_3^{+1}	trig. planar	PF_6^{-1}	octahedral
BF_3	trig. planar	$C_7H_7^{+1}$	heptagonal planar[a])
BrF_3	T shaped	IF_7	pentagonal bipyramidal
NH_3	pyramidal	$C_8H_8^{-2}$	octagonal planar[a])
P_4	tetrahedral	C_8H_8	cubic
Te_4^{+2}	sq. planar	CoF_6^{-3}	octahedral
C_4H_4	sq. planar[a])	$Fe(C_5H_5)_2$	pseudo octahedral
CH_4	tetrahedral	Mo_6Cl_8	octahedral cluster[b])
PO_4^{-3}	tetrahedral	$Nb_6Cl_{12}^{+2}$	octahedral cluster[b])

[a]) Treatment of the π molecular orbitals
[b]) Treatment of metal cluster molecular orbitals.

mechanics of this approach and then return to the theoretical justification at the conclusion. The following abbreviations will be used: BMO = bonding MO, ABMO = antibonding MO, NBMO = non-bonding MO, SO = symmetry orbital, GO = generator orbital. Throughout this article delocalized MO's will be denoted by the symbol ψ and and localized MO's by ϕ.

2. Delocalized MO's of N_2

The nitrogen 2s and 2p valence AO's conveniently oriented for interaction, are shown in Fig. 5. This orientation naturally divides the valence AO's into three orbital sets, namely, the 2s set, the 2 p*x* set and the 2 p*y*, 2 p*z* set. The latter two sets we refer to as radial and tangential p orbitals, respectively. The radial orbitals point toward the molecular center and the tangential orbitals are perpendicular to the radial orbitals. Let us consider these sets separately and imagine various GO's placed one by one between the nitrogen atoms. For each GO we try to find that combination of nitrogen orbitals which has the same symmetry, until we have as many symmetry orbitals (SO's) as there are nitrogen valence orbitals in the set.

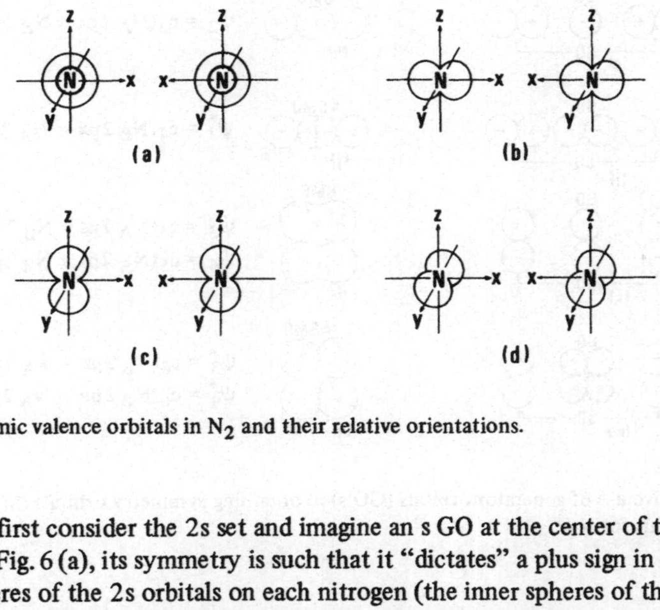

Fig. 5. Atomic valence orbitals in N_2 and their relative orientations.

If we first consider the 2s set and imagine an s GO at the center of the N_2 molecule as in Fig. 6 (a), its symmetry is such that it "dictates" a plus sign in the proximal outer spheres of the 2s orbitals on each nitrogen (the inner spheres of these two orbitals must then be negative since the wave function changes sign across a nodal surface). If we mentally remove the central canonical GO, we are left with a symmetry orbital (SO) composed of two nitrogen 2s AO's. To obtain the second SO from N_A 2s and N_B 2s we examine the symmetry characteristics of additional canonical GO's. From Figs. 6(c), (e) and (f) we see that the p*x* GO generates an SO which has a nodal plane and the other two p GO orbitals do not generate any SO's. A point worth making at this time is that the use of any of the higher canonical GO's would not generate a new SO between the N 2s AO's. Instead, it would either reproduce one of the two SO's just discussed or no SO at all.

It is seen from Figs. 6 (g) and (i) that the s and p*x* GO's also generate the two SO's in the radial p orbital set. On the other hand, SO's for the tangential set are generated

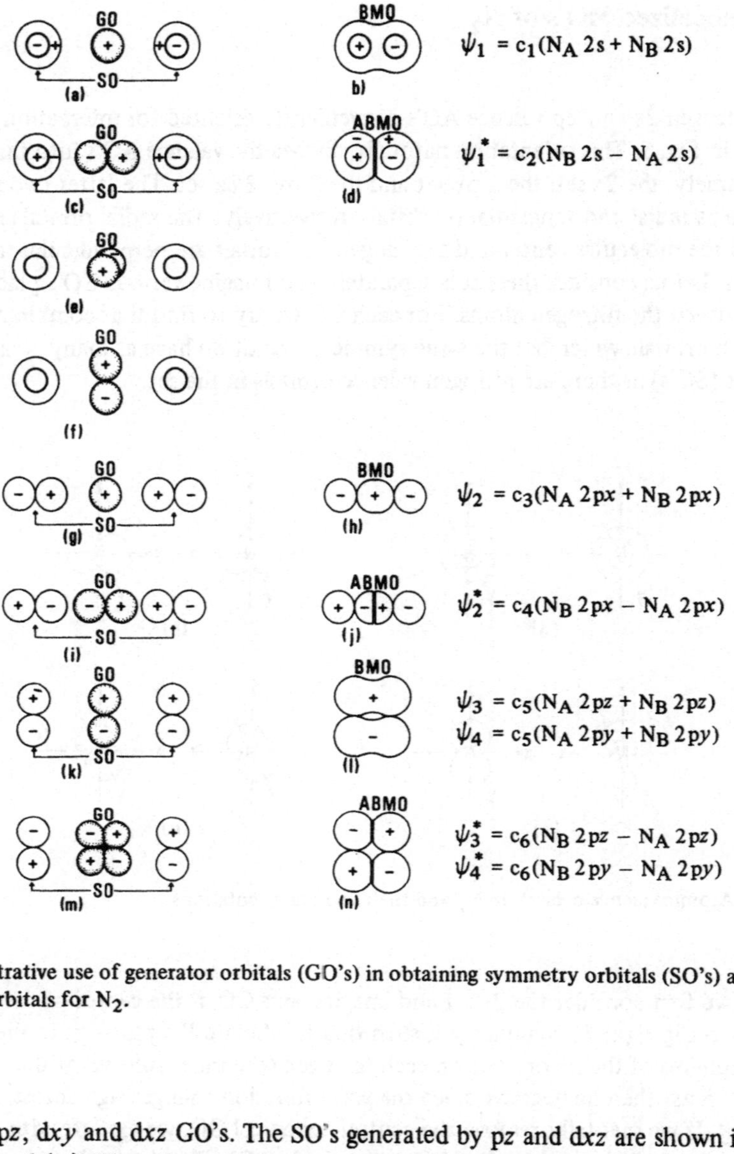

Fig. 6. Illustrative use of generator orbitals (GO's) in obtaining symmetry orbitals (SO's) and molecular orbitals for N_2.

by the py, pz, dxy and dxz GO's. The SO's generated by pz and dxz are shown in Figs. 6 (k) and (m). The other two are perpendicular to these SO's.

Molecular orbitals are formed by combining SO's generated by the same GO's. Thus we begin by taking linear combinations of the SO's generated by the s GO given in Fig. 6 (a) and (g). In general, a linear combination of two SO's will lead to two MO's whose orbital energies are "split apart" (i.e., the higher MO energy lies above both SO energies and the lower MO energy lies below them). However, if the SO energy difference is large then the resulting MO's differ very little from the SO of closest energy. This situation obtains in the case at hand because energetically the nitrogen 2s lies substantially below the nitrogen 2p orbital. Consequently the SO of

Fig. 6 (a) approximately becomes the BMO of Fig. 6 (b), and the SO of Figure 6 (g) approximately becomes the BMO of Fig. 6 (h). If this approximation is not made, one sp mixing coefficient, which can only be determined by numerical calculation, will appear in each delocalized MO. This implies not only that orbital mixing occurs between two σ bonding SO's but it also demonstrates that in general delocalized MO's consist of hybrid valence AO's. Again ignoring this sp mixing, the SO's of Figs. 6 (c) and (i), which are both generated by the px GO, become the ABMO's of Figs. 6 (d) and (j), respectively. Since the py, pz, dxy and dxz GO's generate SO's only in the tangential set, these SO's are not linearly combined with others and hence become a pair of pi BMO's and a pair of pi ABMO's. One of these BMO's and one of these ABMO's are depicted in Fig. 6 (l) and (n); the other two are perpendicular to those depicted.

The qualitative MO energy level diagram in Fig. 7 can be inferred from the rule that there is an approximately equal splitting of the BMO's and their ABMO counterparts and that greater splittings are generally expected from σ than from π interactions.

This procedure may appear to be a rather tedious way of simply taking symmetric and antisymmetric linear combinations of the appropriate pairs of AO's. Even for only slightly more complicated geometries, however, the permitted linear AO combinations are virtually impossible to formulate by inspection only. The generator orbital approach pictorially circumvents the need for using group theory while simultaneously demonstrating that symmetry characteristics of MO's can be directly related to those of AO's. The nodal characteristics of a GO are carried over into the generated SO and hence into the MO. This is clearly seen in Fig. 6.

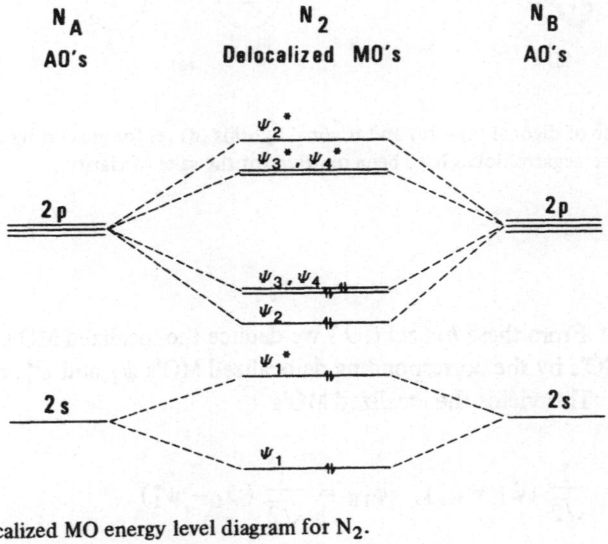

Fig. 7. Delocalized MO energy level diagram for N_2.

3. Localized MO's of N_2

A localized MO view of nitrogen can be arrived at with the help of GO's obtained by hybridizing those canonical GO's which gave rise to the *occupied delocalized* MO's. The lowest two MO's ψ_1 and ψ_1^* in Fig. 7 were generated by an s and a px GO, respectively. We hybridize these two to yield the digonal hybrids

$$h_+ = \frac{1}{\sqrt{2}}(s + px), \quad h_- = \frac{1}{\sqrt{2}}(s - px),$$

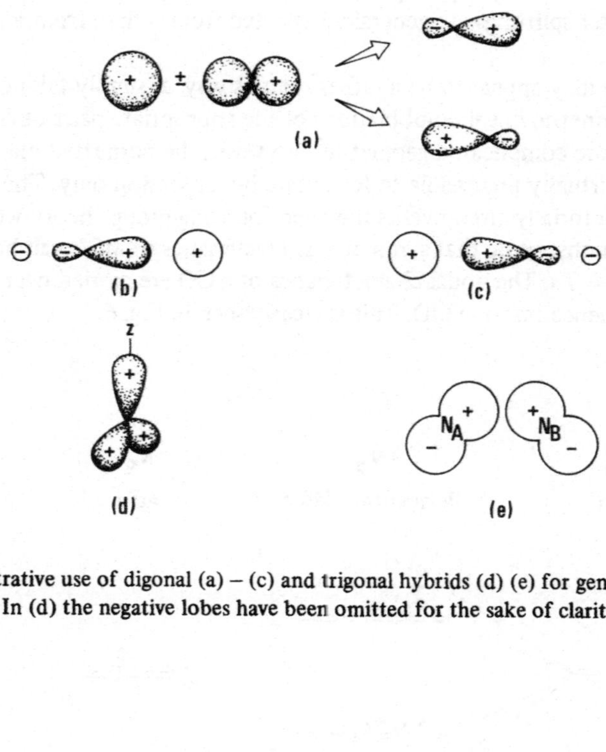

Fig. 8. Illustrative use of digonal (a) – (c) and trigonal hybrids (d) (e) for generating localized MO's in N_2. In (d) the negative lobes have been omitted for the sake of clarity.

shown in Fig. 8 (a). From these *hybrid GO's* we deduce the localized MO's by replacing the s and px GO's by the corresponding delocalized MO's ψ_1 and ψ_1^*, respectively, in these equations. This yields the localized MO's

$$\phi_{1A} = \frac{1}{\sqrt{2}}(\psi_1 + \psi_1^*), \quad \phi_{1B} = \frac{1}{\sqrt{2}}(\psi_1 - \psi_1^*).$$

By inserting the explicit forms for ψ_1 and ψ_1^* we obtain

$$\phi_{1A} = \frac{1}{\sqrt{2}} (c_1 + c_2)(N_A 2s) + \frac{1}{\sqrt{2}} (c_1 - c_2)(N_B 2s)$$

$$\phi_{1B} = \frac{1}{\sqrt{2}} (c_1 + c_2)(N_B 2s) + \frac{1}{\sqrt{2}} (c_1 - c_2)(N_A 2s)$$

and they are shown in Fig. 8(b) and 8(c) along with the corresponding hybrid GO's. Since $c_1 = (2 + 2S)^{-1/2}$ and $c_2 = (2 - 2S)^{-1/2}$ where S is the overlap integral between N_A 2s and N_B 2s, it is apparent that $c_1 - c_2$ is considerably smaller than $c_1 + c_2$. Consequently ϕ_{1A} is largely centered on atom N_A and ϕ_{1B} on N_B, and this is indicated in Fig. 8(b) and 8(c). Orthogonality between the two localized MO's furthermore requires that the minor contribution is opposite in sign to the major contribution, and this is in agreement with the fact that c_2 is greater than c_1. Thus the two electrons in each of the localized MO's have strong lone pair character and the cancellation of the bonding and antibonding effects of ψ_1 and ψ_1^* can be approximately described by placing the four electrons in two atomic-like lone pair orbitals. Anticipating this situation, we could have reserved the 2s orbitals for the lone pairs and subsequently omitted them from the bonding description. In more complicated molecules, this will be a convenient procedure.

The localization procedure just outlined can also be applied to the ls AO's. Their delocalized description leads naturally to a very slightly binding and antibinding MO owing to the extremely small overlap. Localizing them gives an inner shell orbital centered on each nitrogen. Due to the small overlap, these inner shell orbitals are practically pure one-center 1s AO's. The one-center character of these and inner shells in this and other molecules is tacitly assumed in discussions where valence orbitals are used exclusively.

The upper three bonding MO's in Fig. 7 were generated by s, p*y* and p*z* canonical GO's. Making hybrid GO's from these orbitals, we produce the trigonal set shown in Fig. 8(d). The one pointing in the *z* direction is given by the formula

$$\text{tr1} = \frac{1}{\sqrt{3}}\, s + \sqrt{\frac{2}{3}}\, pz\,.$$

Substituting in this equation the bonding MO's ψ_2 and ψ_3 for the GO's s and pz, respectively, we obtain the localized bonding MO

$$\phi_1 = \frac{1}{\sqrt{3}}\, \psi_2 + \sqrt{\frac{2}{3}}\, \psi_3$$

If we now insert the explicit expressions for ψ_2 and ψ_3 we find

$$\phi_1 = c_3 \frac{1}{\sqrt{3}} (N_A\,2px) + c_5 \sqrt{\frac{2}{3}} (N_A\,2pz) + c_3 \frac{1}{\sqrt{3}} (N_B\,2px) + $$
$$+ c_5 \sqrt{\frac{2}{3}} (N_B\,2pz)$$
$$= c_7(N_A\,2p\hat{x} + N_B\,2p\hat{x}),$$

where

$$2p\hat{x} = (c_3/c_7\sqrt{3})\,2px + (c_5\sqrt{2}/c_7\sqrt{3})\,2pz$$

represents a p orbital, pointing in a direction $\hat{x}$ intermediate between $2px$ and $2pz$. It is thus apparent that ϕ_1 is made up of two canted p orbitals ($N_A\,2p\hat{x}$ and $N_B\,2p\hat{x}$) as indicated in Fig. 8 (e), i.e., it is a "bent bond" in the region toward which the sp^2 GO pointed. Since the sp^2 GO's are trigonally disposed, the other two sp^2 hybrid GO's generate localized bent bonds which can be obtained from the one just discussed by a rotation of 120° around the bond axis. Thus one sigma and two pi bonds can be recast into three banana bonds.

From the foregoing it is apparent that the 3 banana bonds are "induced" by the directionality of the sp^2 GO's. By the same token the directionality of the sp^2 GO's can be considered to "induce" the *contributions* to any of the three banana bonds on each of the N atoms. These contributions are, however, linear combinations of N valence p AO's and hence themselves are p AO's rotated in new directions. Thus the directionality of the sp^2 GO's can be imagined to induce a set of three p orbitals $2p\hat{x}$,

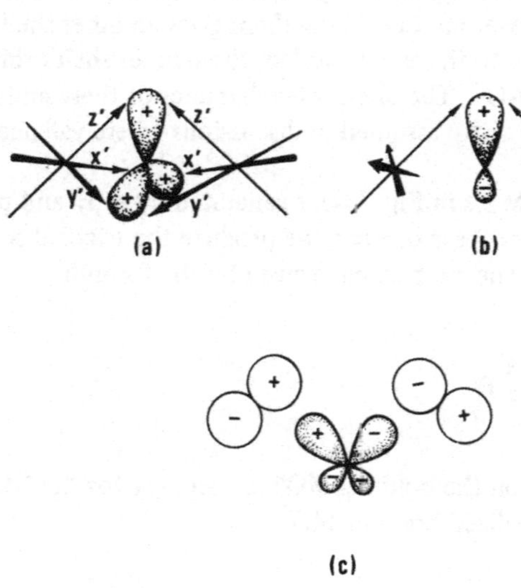

Fig. 9. Illustrative use of trigonal hybrids GO's in generating localized BMO's and ABMO's in N_2 using nitrogen p orbital sets which are directed toward one another. In (a) and (b) only the axes of these p AO's are shown, and their direction represents the direction of the positive lobe. In (c) is shown one member of the pd^2 hybrid GO set which generates the localized ABMO's.

$2p\hat{y}$, $2p\hat{z}$ on each N which point in the bond direction in a trigonally disposed manner. An appropriately rotated set of *orthogonal* p AO's as depicted in Figure 9 (a) would have the proper orientation. However the p AO's generated are in fact not orthogonal because in general the coefficients c_3 and c_5 are not equal. If c_3 is greater (smaller) than c_5, these p orbitals are tilted towards (away) from the bond axis. Thus we can imagine each of the banana bonds as resulting from a bonding combination of two appropriately oriented canonical p AO's on each nitrogen:

$$\phi_1 = c_7(N_A\,2p\hat{x} + N_B\,2p\hat{x})$$
$$\phi_2 = c_7(N_A\,2p\hat{y} + N_B\,2p\hat{y})$$
$$\phi_3 = c_7(N_A\,2p\hat{z} + N_B\,2p\hat{z})$$

where the p orbitals on a given nitrogen are orthogonal if $c_3 = c_5$. Thus we see that the atomic contributions to localized MO's can be inferred from the appropriate hybrid GO's. It is apparent that the Lewis structure can be considered as a schematic representation of the occupied, localized MO structure if each dash stands for a localized orbital.

The three antibonding MO's ψ_2^*, ψ_3^*, ψ_4^* in Fig. 7 can be localized in an analogous fashion with the help of trigonal GO's obtained by hybridizing the canonical GO's which generated these delocalized ABMO's. The hybrid GO pointed in the *z* direction is given by

$$\mathrm{tr}1^* = \frac{1}{\sqrt{3}}\,px + \sqrt{\frac{2}{3}}\,dxz$$

and is schematically shown in Fig. 9 (c) together with the generated localized ABMO. Following the preceding reasoning we find for the three antibonding banana bonds the linear combinations

$$\phi_1^* = c_8(N_A\,2px' - N_B\,2px')$$
$$\phi_2^* = c_8(N_A\,2py' - N_B\,2py')$$
$$\phi_3^* = c_8(N_A\,2pz' - N_B\,2pz'),$$

where $2px'$, $2py'$, $2pz'$ are appropriately rotated p AO's.[4])

The corresponding MO level diagram is shown in Fig. 10. It must be clearly recognized, however; that this energy level diagram has resulted from localizing separately the ground state occupied and unoccupied delocalized MO's. A different molecular state will have different orbital occupations, and hence will lead to different localized MO's and different localized energy levels.

---

[4]) If $c_3 = c_5$, $2p\hat{x} = 2px'$ $2p\hat{y} = 2py'$ and $2p\hat{z} = 2pz'$ and these AO's are mutually orthogonal. If $2p\hat{x}$, $2p\hat{y}$ and $2p\hat{z}$ are tilted toward the bond axis (i.e., the angles between them are less than 90°) then $2px'$, $2py'$ and $2pz'$ are tilted away from the bond axis, and *vice versa*.

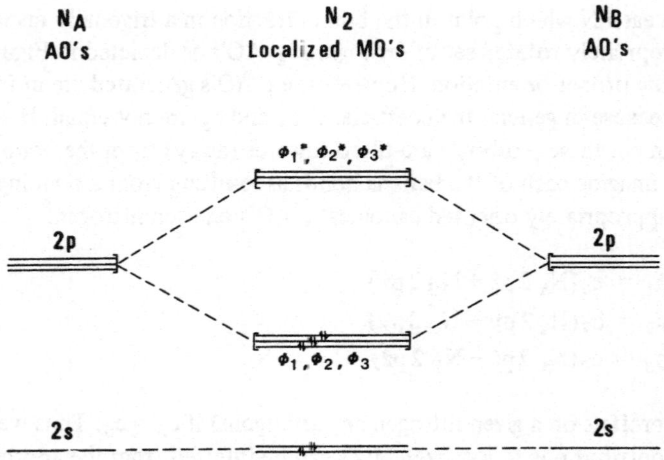

Fig. 10. Localized energy level diagram for N_2.

Because the s GO generates the BMO ψ_1 from the N 2s AO's and also the BMO ψ_2 from the N 2px AO's, some mixing of 2s and 2px character occurs in the actual delocalized MO's, as mentioned earlier. The same is true for the ABMO's ψ_1^* and ψ_2^* and also for the localized MO's. In the latter, this mixing leads to an sp hybridization which places some p character in the lone pairs and some s character in the banana bonds as depicted below. Even so the lone pair orbitals have substantially more s character than the bonding orbitals, and this leads to the VSEPR idea that the lone pair orbitals occupy more space around atoms then do bond pairs. Consequently the

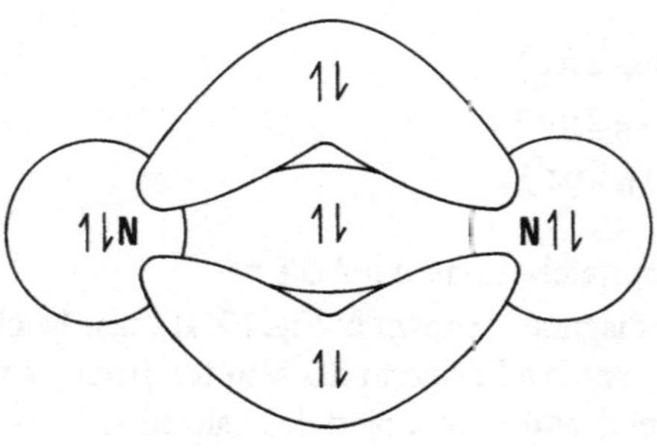

bonding orbitals of N_2 form smaller angles with the bond axis than would sp^3 hybrids. Thus Lewis as well as VSEPR concepts naturally emanate from the localized MO view. In more complicated cases Lewis structures and VSEPR rules can be used to anticipate certain aspects of the MO structure which could only be obtained by explicit calculations.

4. Delocalized MO's for BH_3

The trigonal planarity of BH_3 is easily deduced from VSEPR considerations of the dispositions of the three bonding electron pairs. The *delocalized*

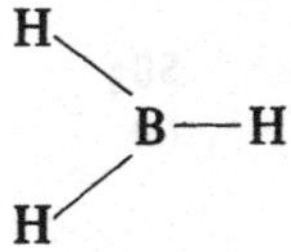

MO bonding view of this molecule can be derived by considering the interactions of the hydrogen SO's with boron canonical valence AO's. In Fig. 11 (a)–(c) are shown the three permitted hydrogen SO's which are generated by central canonical GO's[5] (the px GO is pointed toward one hydrogen). The SO's on the boron are simply the canonical valence orbitals and are trivially generated by the GO's. In the present case we have constructed two sets of SO's, the hydrogen SO's and the boron valence AO's. The BH_3 MO's are obtained by combining, in a bonding and antibonding fashion,[6] SO's which arise from the *same* GO's as shown in Fig. 12. The preconstruction of different sets of SO's is always convenient when the molecule contains sets of equivalent valence AO's.

The directions of the GO's with respect to the hydrogens is, however, arbitrary. For instance we can rotate them by 45° so that the px and py GO's are equivalently positioned with respect to one of the hydrogens as shown in Fig. 11 (d) and (e). It is apparent that here the two SO's will be mirror images of each other and hence degenerate. From this it can be inferred that such is also the case for SO_2 and SO_3, even though this degeneracy is not immediately apparent. It is often the case that certain canonical GO's having the same l value produce degenerate MO's and this can usually be recognized by appropriate positioning of the GO's. It may be noted that there is a tendency among novices to postulate MO's constructed by simply permuting the signs on the three H1s AO's. The GO procedure avoids the problems arising from such an approach.

---

[5] The coefficients of the H 1s AO's in a given SO are not necessarily equal. For SO_3 one of the orbital coefficients is seen to be zero from Fig. 11 (c) while the other two hydrogen AO's have coefficients of identical magnitude but opposite sign. For SO_2 the values of the coefficients are easily deduced by demanding orthogonality with SO_1. Thus the 1s contribution for one of the hydrogens must be of twice the magnitude and opposite in sign to the coefficient of the other two hydrogen contributions in order to have orbital sign "balance".

[6] In the case of N_2, the two s-GO-generated SO's were both bonding and their linear combination merely produced modified BMO's. In BH_3, the SO's in themselves are essentially non bonding because of the large interatomic separation of the hydrogens. Linear combination of SO's from different sets of atoms will yield BMO's and ABMO's if the atom sets are close enough to interact.

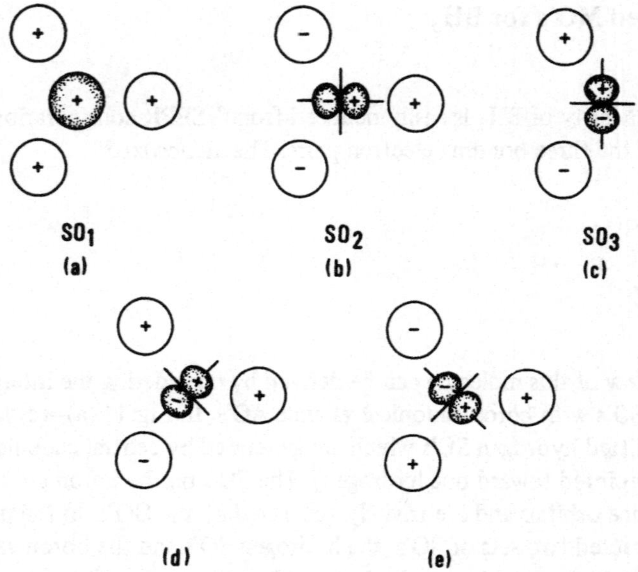

Fig. 11. Hydrogen symmetry orbitals for BH_3 generated by two different axis systems ((a)–(c) and (d) (e)).

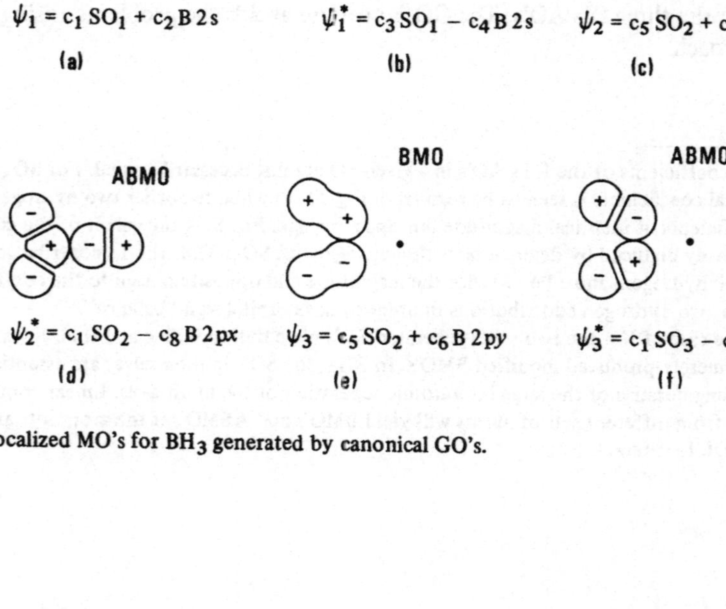

Fig. 12. Delocalized MO's for BH_3 generated by canonical GO's.

The energy level diagram shown in Fig. 13 along with the six electrons from the Lewis structure shown earlier shows that the lower three BMO's are filled. These MO's are four-center two-electron bonds and are therefore delocalized over the whole molecule.

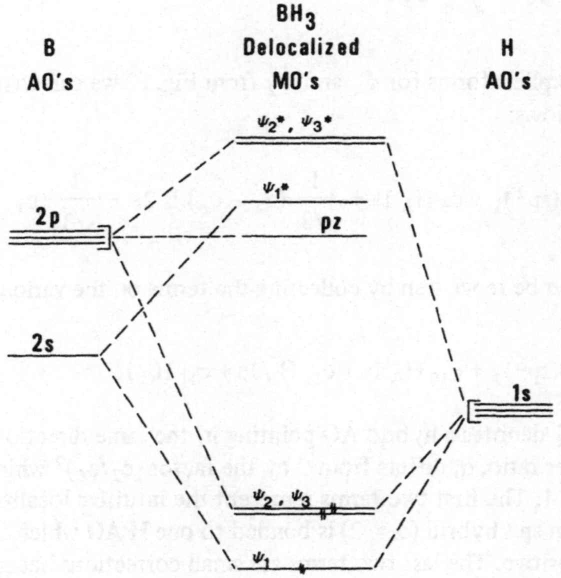

Fig. 13. Delocalized MO energy level diagram for BH_3.

5. Localized MO's of BH_3

In order to obtain a localized view for BH_3, we again use appropriate hybrid GO's made up from those canonical GO's which generated the *occupied* delocalized MO's, In the case at hand these are sp^2 trigonal hybrids and the most effective localized arrangement is obtained by pointing the three hybrid GO's toward the three hydrogens as shown in Fig. 14 (a). In this way two-center two-electron bonds will result, in accord with the Lewis structure.

If the x axis points from B to H_A then the corresponding hybrid GO is

$$h_x = \frac{1}{\sqrt{3}}\, s + \sqrt{\frac{2}{3}}\, px\,.$$

As before we deduce a corresponding localized MO through replacing the canonical GO's s and p, respectively, by the delocalized MO's which they generated, namely, ψ_1 and ψ_2, respectively, in Fig. 12. Thus we obtain

$$\phi_1 = \frac{1}{\sqrt{3}} \psi_1 + \sqrt{\frac{2}{3}} \psi_2 .$$

By inserting the explicit forms for ψ_1 and ψ_2 from Fig. 12 we can write this localized orbital as follows:

$$\phi_1 = c_6 \, B(sp^2)_1 + c_5 \, H_A \, 1s + \frac{1}{\sqrt{3}} (c_2 - c_6) \, B \, 2s + \frac{1}{\sqrt{3}} (c_1 - c_5) \, SO_1 .$$

This expression can be re-written by collecting the terms on the various atoms to give

$$\phi_1 = c_9 \, B(sp^\alpha)_1 + c_{10} \, H_A \, 1s + c_{11} \, H_B \, 1s + c_{11} \, H_C \, 1s .$$

The symbol $(sp^\alpha)_1$ denotes a hybrid AO pointing in the same direction as $(sp^2)_1$ but whose s/p character ratio, α, differs from 2 by the factor $(c_2/c_6)^2$ which is slightly different from 1. The first two terms represent the intuitive localized picture, namely, that one boron sp^α hybrid ($\alpha \simeq 2$) is bonded to one H AO which implies that c_9 and c_{10} are both positive. The last two terms are small corrections because $c_{11} = (c_1 - c_5)/\sqrt{3}$; orthogonality requires furthermore that they be negative. These corrections could have been anticipated from the placement and sizes of the plus and minus lobes of the hybrid GO as shown in Fig. 14 (b). If these correction terms are neglected, we arrive at the conventional picture of localized bonds in terms of hybrid AO's.

The other two localized MO's are obtained from the other two hybrid GO's by the same procedure. More simply they can be obtained from the one just discussed

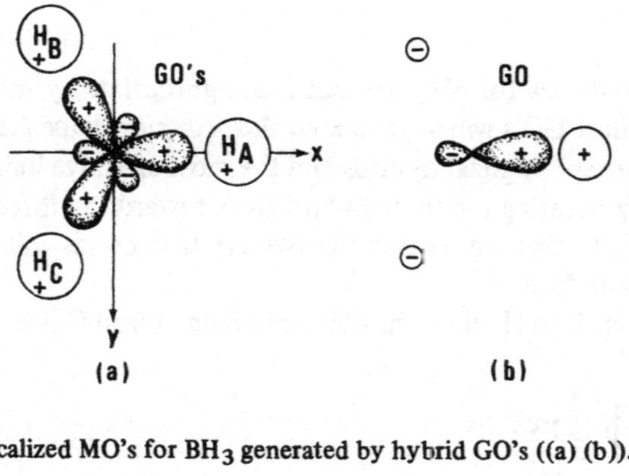

Fig. 14. Localized MO's for BH_3 generated by hybrid GO's ((a) (b)).

by a rotation of 120°. This follows from the disposition of the hybrid GO's. The B sp$^{\alpha}$ contributions to the three localized MO's induced by the sp^2 hybrid GO's are also trigonally arranged even though α is not exactly 2, which implies that $B(sp^{\alpha})_{1,2,3}$ are not orthogonal. The three approximate localized bonding MO's with appropriate expressions are shown below.

$$\phi_1 = c_1 B(sp^{\sim 2})_1 + c_2 H_A 1s$$
$$\phi_2 = c_1 B(sp^{\sim 2})_2 + c_2 H_B 1s$$
$$\phi_3 = c_1 B(sp^{\sim 2})_3 + c_2 H_C 1s$$

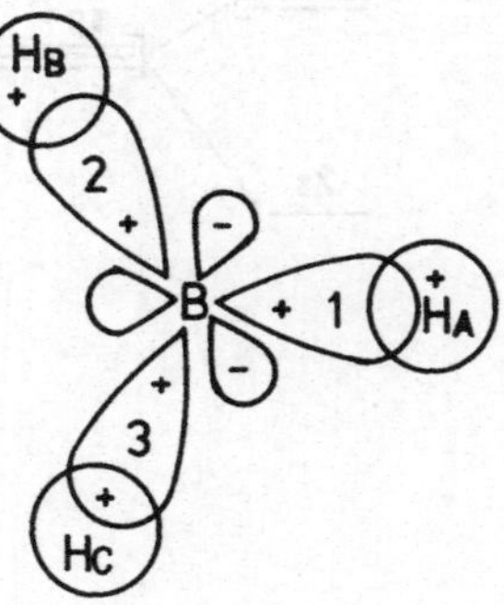

From the delocalized ABMO's one can form localized ABMO's in an analogous fashion and in this case using the same GO's. This will yield the following expressions:

$$\phi_1^* = c_3 B(sp^{\sim 2})_1 - c_4 H_A 1s$$
$$\phi_2^* = c_3 B(sp^{\sim 2})_2 - c_4 H_B 1s$$
$$\phi_3^* = c_3 B(sp^{\sim 2})_3 - c_4 H_C 1s .$$

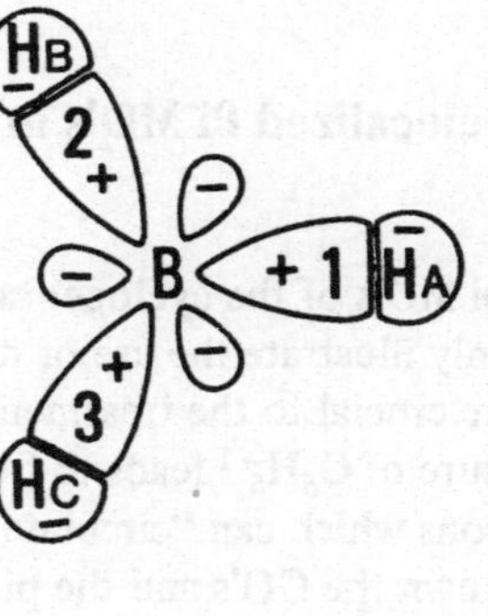

The energy diagram for the localized BMO's and ABMO's is given in Fig. 15. The three electron pairs are seen to be localized in three equivalent two-center two-electron BMO's and it is this localized description which is symbolized by the Lewis structure.

The case just treated exemplifies the general phenomenon that the AO contributions of the central atom to the localized bonds are very close to the corresponding GO hybrids. This justifies the common practice of assuming central atom valence orbital hybridization for a localized bonding picture. In addition the GO approach also shows that the hybrid mixing ratio of the valence AO's generally deviates somewhat from that of the GO hybrids while preserving their appropriate directional disposition. Inferences concerning GO-induced hybridizations on non-central atoms can also be deduced.

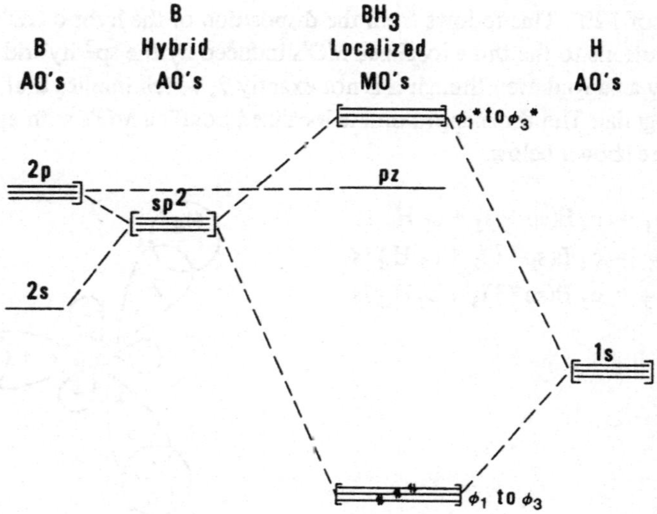

Fig. 15. Localized MO energy level diagram for BH_3.

6. Delocalized Pi MO's in $C_5H_5^{-1}$

The pi MO's of the cyclopentadienide ion are instructive to consider because they not only illustrate the use of d and f GO's, but the bonding considerations in this ion are crucial to the treatment of ferrocene, our next example. The electron dash structure of $C_5H_5^{-1}$ leads to five Lewis structures suggesting that there are six pi electrons which can "circulate around the ring" in the delocalized view. In Fig. 16 are shown the GO's and the pi SO's generated by them. Because there are no other atoms, the five SO's become the five MO's expected because there are five carbon AO's available for pi bonding.

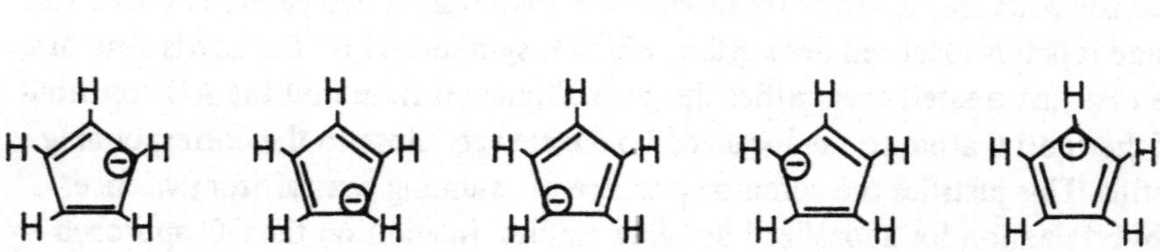

The d and f GO's give rise to two degenerate pairs of delocalized MO's. As in the case of the delocalized picture of BH_3, this degeneracy can be made apparent by

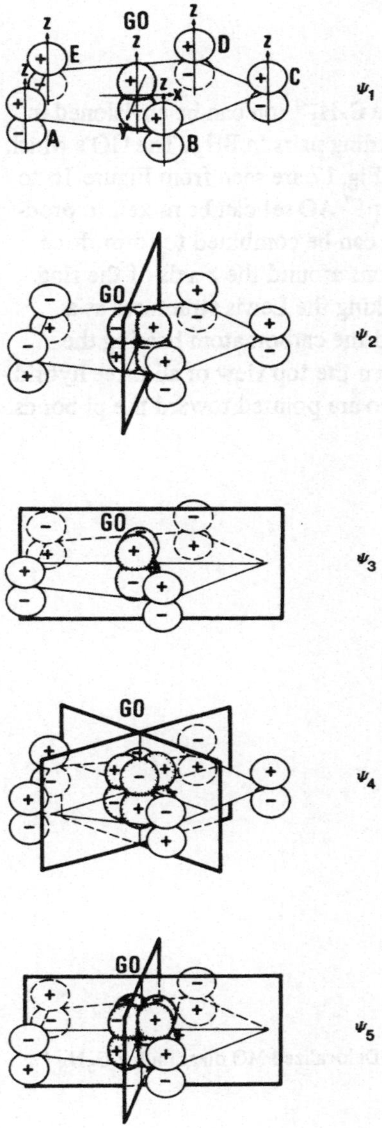

Fig. 16. Generation of the symmetry orbitals in $C_5H_5^{-1}$ by means of canonical GO's. The node(s) perpendicular to the ring are shown using heavy lines.

proper positioning of the GO's. The ordering of orbital energies in Figure 17 can be qualitatively inferred from the number of nodes between neighboring atoms in each MO. Since ψ_1 has no such nodes, it is a totally bonding MO, i.e., it has bonding interactions between each pair of neighboring AO's. In ψ_2, there are three such bonding interactions *versus* two anti-bonding ones, leaving an excess of one bonding interaction. The same holds for ψ_3 with two bonding and one antibonding interaction. Similar reasoning leads to the conclusion that ψ_4 and ψ_5 are antibonding.

7. Localized Pi MO's in $C_5H_5^{-1}$

The localized bonding of the six pi electrons in the $C_5H_5^{-1}$ ion can be envisioned in much the same way as was done for the sigma bonding pairs in BH_3. The GO's which lead to the occupied delocalized bonding MO's in Fig. 17 are seen from Figure 16 to be p0, d1′ and d1″. In the same way that an s, p1′, p1″ AO set can be mixed to produce three equivalent sp^2 hybrids, a p0, d1′, d1″ set can be combined to form three equivalent pd^2 hybrids pointed in trigonal directions around the *z* axis of the ring. One of these hybrids is shown in Figure 18 (a). Taking the Lewis structures as a reference, we point one of the hybrid GO's toward the carbon atom bearing the formal negative charge as shown in Fig. 18 (b). From the top view of all three hybrid GO's in Fig. 18 (c), it is apparent that the other two are pointed toward the pi bonds.

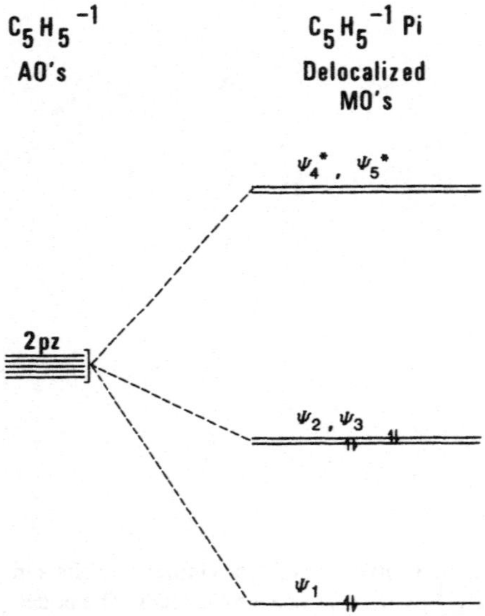

Fig. 17. Delocalized MO diagram for $C_5H_5^{-1}$.

Using these hybrid GO's in the same way as in BH_3, we obtain the localized pi MO's schematically depicted in Fig. 19 (a) (b), each of which houses a pair of electrons. The top view shows that two of these localized MO's are equivalently positioned with respect to atom centers and consequently have the same shape. The third localized MO, being differently placed with respect to atom centers, has a different shape. The Figure shows that the Lewis structure, though qualitatively correct, is an oversimplification in that the unique localized MO is in fact almost a 3-center bond and the double bonds are correspondingly polarized.

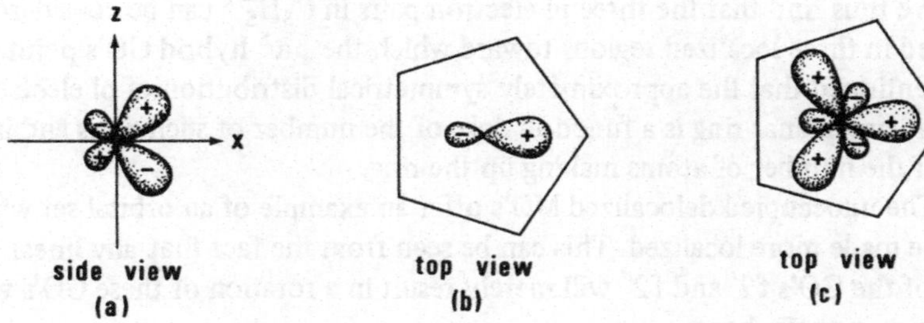

Fig. 18. Illustrative use of p d² hybrid GO's to generate localized MO's of $C_5H_5^{-1}$.

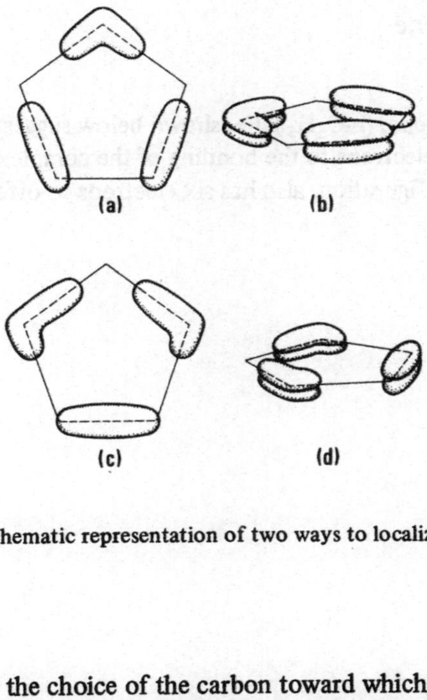

Fig. 19. Schematic representation of two ways to localize the bonding pi MO electron pairs in $C_5H_5^{-1}$.

Since the choice of the carbon toward which we pointed the first hybrid GO is arbitrary, there are five different localized MO *sets*. Contrary to valence bond usage, these five MO representations are not to be superimposed (as resonance structures would be). Rather, they are all equivalent and each by itself is an adequate representation of the molecular state.

In contrast to BH_3, another useful localized description is obtained in this case from a different orientation of the hybrid GO's. Pointing one of these hybrid GO's toward the midpoint of a bond gives rise to the localized MO's shown in Fig. 19 (c) and (d). It is apparent that while all these GO's point between carbon atom pairs, two are off-center.

We thus find that the three pi electron pairs in $C_5H_5^{-1}$ can be considered as housed in three localized regions toward which the pd^2 hybrid GO's point. It may be mentioned that the approximately symmetrical distribution of pi electron pairs around any planar ring is a function only of the number of such pairs and is independent of the number of atoms making up the ring.

The unoccupied delocalized MO's offer an example of an orbital set which can not be made more localized. This can be seen from the fact that any linear combination of the GO's f2′ and f2″ will merely result in a rotation of these GO's without a change in their shape.

8. Delocalized MO's of Ferrocene

The electron dash structure of ferrocene, $(h\text{-}C_5H_5)_2Fe$, shown below suggests that each ligand anion contributes six pi electrons to the bonding of the complex while the Fe^{+2} ion, having an $Ar(3d)^6$ configuration, also has six electrons to offer from

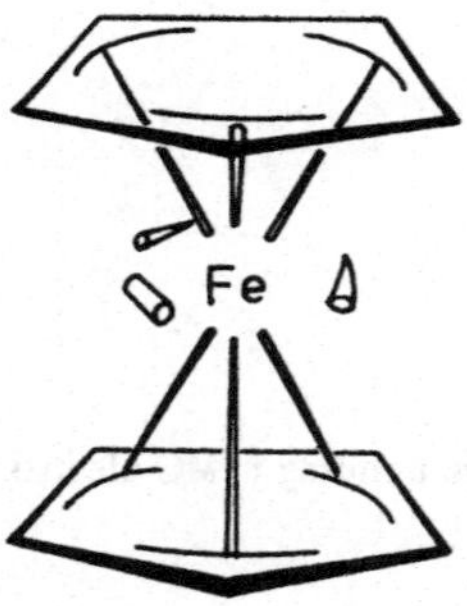

its valence shell. To develop the delocalized view we begin by noting that each $C_5H_5^{-1}$ ion contributes five carbon pz AO's (perpendicular to the plane of the ring) to the ferrocene MO's. We place GO's at the central position between the two rings (i.e. in the position of the metal atom) in order to determine the permitted SO's formed by the two $C_5H_5^{-1}$ pz AO sets as shown schematically in Fig. 20. Each of the ten SO's shown for the two $C_5H_5^{-1}$ rings is seen to contain, in each separate $C_5H_5^{-1}$ ring, one of the five MO's generated in the previous section. Therefore we show the ligand MO energies in the right hand column of Figure 21, and they are clearly identical with those shown in Fig. 17 for a single $C_5H_5^{-1}$ ion.

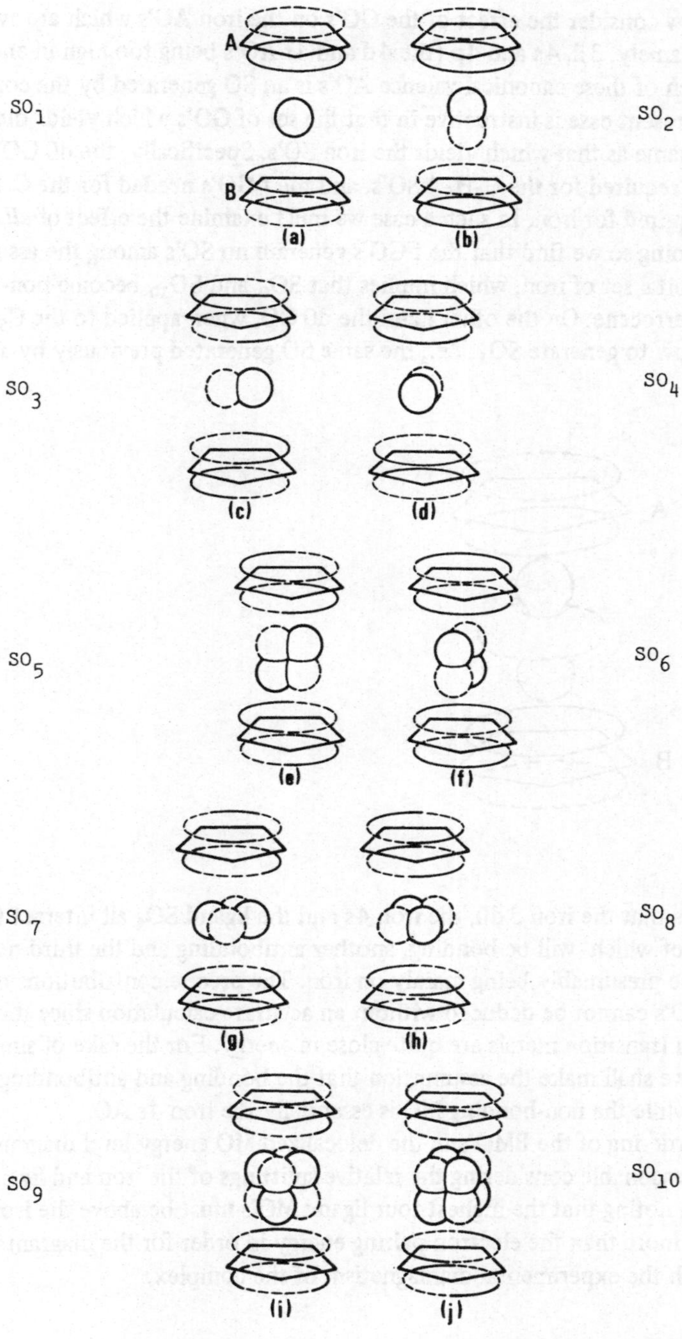

Fig. 20. Generation of the ten symmetry orbitals for ferrocene using generator atomic orbitals. Full and dotted lines in the $C_5H_5^{-1}$ MO's and in the GO's represent opposite signs in the wave functions.

We now consider the effect of the GO's on the iron AO's which are available for bonding, namely, 3d, 4s and 4p (the 4d and 4f AO's being too high in energy). Clearly each of these canonical valence AO's is an SO generated by the corresponding GO. The present case is instructive in that the set of GO's which yields the $C_5H_5^{-1}$ SO's is not the same as that which yields the iron SO's. Specifically, the d0 GO needed for iron is not required for the $C_5H_5^{-1}$ SO's, and the f GO's needed for the $C_5H_5^{-1}$ rings are not required for iron. In such a case we must examine the effect of *all* GO's on *all* AO's. In doing so we find that the f GO's generate no SO's among the assumed valence orbital set of iron, which implies that SO_9 and SO_{10} become non-bonding MO's for ferrocene. On the other hand the d0 GO, when applied to the $C_5H_5^{-1}$ ligands, is seen below to generate SO_1, i.e., the same SO generated previously by an s GO.

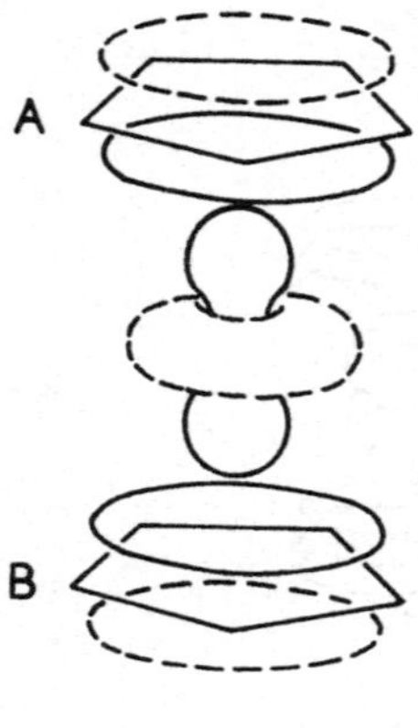

This means that the iron 3 d0, the iron 4s and the ligand SO_1 all interact to give three MO's,one of which will be bonding, another antibonding and the third non-bonding, the last one presumably being mainly on iron. The precise contributions of the constituent SO's cannot be deduced without an accurate calculation since the 3 d0 and 4s levels in transition metals are quite close in energy. For the sake of simplicity, however, we shall make the assumption that the bonding and antibonding MO's contain 3 d0 while the non-bonding MO is essentially the iron 4s AO.

The ordering of the BMO's in the delocalized MO energy level diagram in Figure 21 is reasonable considering the relative splittings of the iron and ligand orbitals. It is worth noting that the highest four ligand MO's must be above the iron 4s in energy by more than the electron pairing energy in order for the diagram to be consistent with the experimental diamagnetism of the complex.

9. Localized MO's of Ferrocene

A satisfactory localized view of ferrocene can be developed by noting from the Lewis structure that there are nine electron pairs around iron; three from each cyclopentadienide ring and three from the metal atom. From our delocalized MO picture we note that the occupied orbitals in Fig. 21 were generated from a set of GO's composed of an s, all three p's and all of the d's. These GO's were seen to be of the same type as the valence shell AO's actually employed by the iron atom in forming MO's. The Lewis structure suggests that we should localize three bonding electron pairs between each ring and the metal atom. To accomplish this we first hybridize the GO's p0, p1′, p1″, d0, d1′, d1″ associated with the lowest six delocalized MO's in Fig. 21 which yields

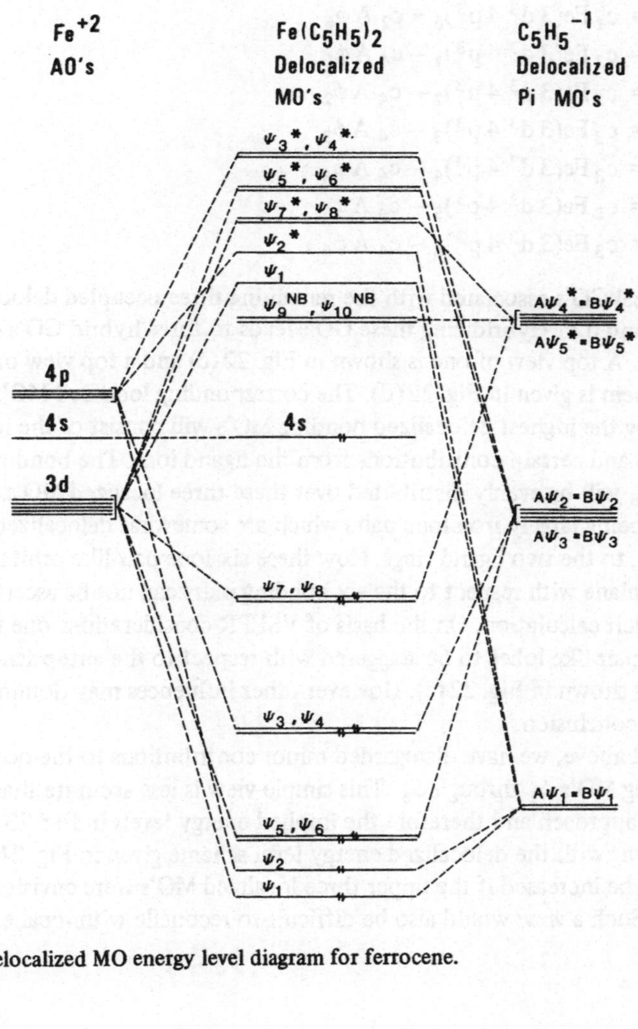

Fig. 21. A delocalized MO energy level diagram for ferrocene.

the trigonal antiprismatically directed p^3d^3 hybrid GO's shown in Figure 22(a). The localized MO's generated by these GO's will essentially consist of the corresponding iron 3 d^3 4 p^3 hybrids and localized ligand orbitals concentrated in those regions of the rings toward which the GO's point (see Figure 22(b)). The ligand contributions are the localized orbitals discussed in the previous section. The same p^3d^3 hybrid GO's can also be used to localize the *unoccupied* delocalized ABMO's generated from the corresponding canonical GO's. Approximate expressions for the resulting localized BMO's and ABMO's are shown below:

$$\phi_1 = c_1 \mathrm{Fe}(3\,d^3\,4\,p^3)_1 + c_2\,A\,\phi_1$$
$$\phi_2 = c_1 \mathrm{Fe}(3\,d^3\,4\,p^3)_2 + c_2\,A\,\phi_2$$
$$\phi_3 = c_1 \mathrm{Fe}(3\,d^3\,4\,p^3)_3 + c_2\,A\,\phi_3$$
$$\phi_4 = c_1 \mathrm{Fe}(3\,d^3\,4\,p^3)_4 + c_2\,A\,\phi_4$$
$$\phi_5 = c_1 \mathrm{Fe}(3\,d^3\,4\,p^3)_5 + c_2\,A\,\phi_5$$
$$\phi_6 = c_1 \mathrm{Fe}(3\,d^3\,4\,p^3)_6 + c_2\,A\,\phi_6$$
$$\phi_1^* = c_3 \mathrm{Fe}(3\,d^3\,4\,p^3)_1 - c_4\,A\,\phi_1$$
$$\phi_2^* = c_3 \mathrm{Fe}(3\,d^3\,4\,p^3)_2 - c_4\,A\,\phi_2$$
$$\phi_3^* = c_3 \mathrm{Fe}(3\,d^3\,4\,p^3)_3 - c_4\,A\,\phi_3$$
$$\phi_4^* = c_3 \mathrm{Fe}(3\,d^3\,4\,p^3)_4 - c_4\,A\,\phi_4$$
$$\phi_5^* = c_3 \mathrm{Fe}(3\,d^3\,4\,p^3)_5 - c_4\,A\,\phi_5$$
$$\phi_6^* = c_3 \mathrm{Fe}(3\,d^3\,4\,p^3)_6 - c_4\,A\,\phi_6\,.$$

The canonical GO's associated with the remaining three occupied delocalized MO's are s, d2′ and d2″. Hybridizing these GO's leads to three hybrid GO's in the equatorial plane. A top view of one is shown in Fig. 22(c) and a top view of the main lobes of all of them is given in Fig. 22(d). The corresponding localized MO's resulting from superposing the highest delocalized bonding MO's will consist of the identical sd^2 iron hybrids and certain contributions from the ligand ions. The bonding character of ψ_7 and ψ_8 will be evenly distributed over these three localized MO's. They can be described as being largely iron lone pairs which are somewhat delocalized, in a bonding fashion, to the two ligand rings. How these six lone pair-like orbitals are oriented in this plane with respect to the six bonding pairs can not be ascertained without an explicit calculation. On the basis of VSEPR considerations one would expect the lone pair-like lobes to be staggered with respect to the anti-prismatic bonding lobes as shown in Fig. 22(e). However other influences may dominate which would alter this conclusion.

As discussed above, we have disregarded minor contributions to the occupied localized bonding MO's ϕ_1 through ϕ_6. This simple view is less accurate than a rigorous localization approach and therefore the implied energy levels in Fig. 23 are not entirely equivalent with the delocalized energy level scheme given in Fig. 21. This deviation would be increased if the upper three localized MO's were envisioned as pure iron lone pairs. Such a view would also be difficult to reconcile with local electroneutrality.

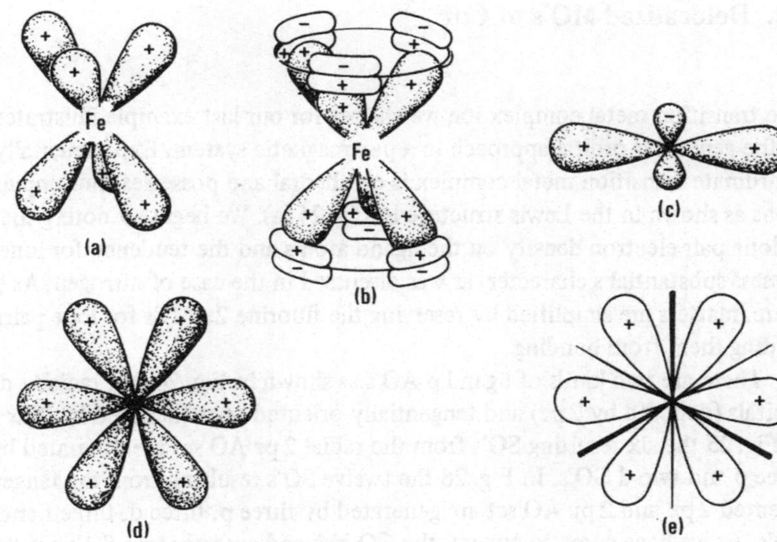

Fig. 22. The use of p^3d^3 (a) (b) and sd^2 (c) (d) hybrid GO's for the localized view of ferrocene. In (e) is shown a top view of the localized orbitals around iron (see text).

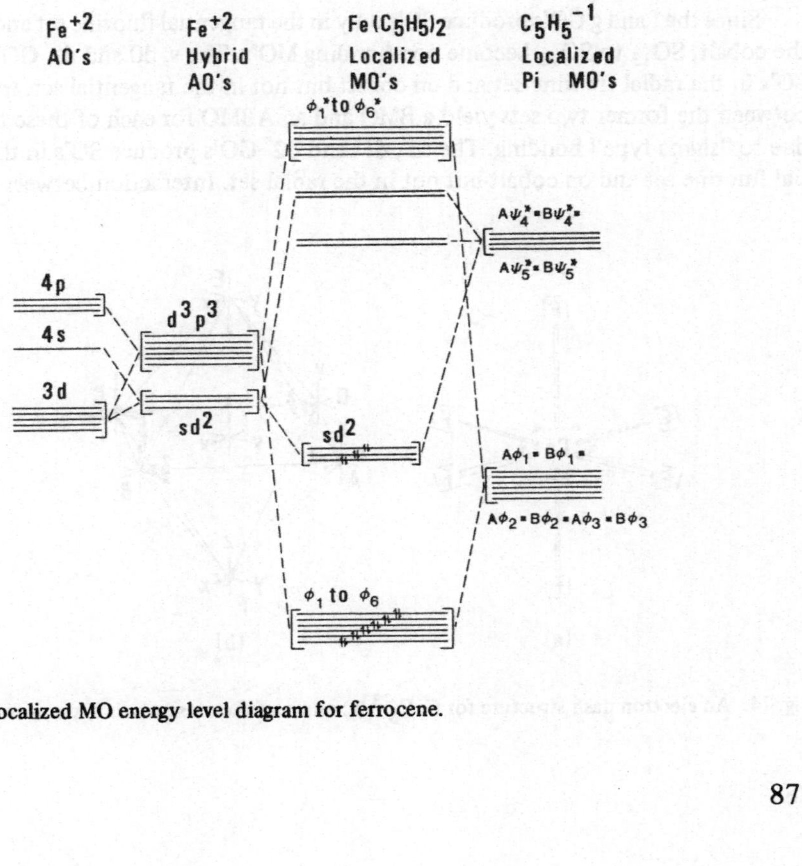

Fig. 23. A localized MO energy level diagram for ferrocene.

10. Delocalized MO's of CoF_6^{-3}

The transition metal complex ion we choose for our last example illustrates the use of the generator orbital approach in a paramagnetic system. Experimentally this six-coordinate transition metal complex is octahedral and possesses four unpaired electrons as shown in the Lewis structure in Fig. 24 (a). We begin by noting the presence of lone pair electron density on the ligand atoms and the tendency for lone pairs to possess substantial s character, as was discussed in the case of nitrogen. As suggested there, matters are simplified by reserving the fluorine 2s AO's for lone pairs and excluding them from bonding.

There are two kinds of ligand p AO's as shown in Fig. 24 (b): radially directed orbitals (denoted by 2*pz*) and tangentially oriented ones (denoted by 2*px* and 2*py*). In Fig. 25 the six resulting SO's from the radial 2*pz* AO set are generated by an s, three p and two d GO's. In Fig. 26 the twelve SO's resulting from the tangentially oriented 2*px* and 2*py* AO set are generated by three p, three d, three f and three g GO's. As we have come to expect, the GO-induced central atom SO's are the energetically accessible valence AO's. They are the cobalt 3d, 4s and 4p AO's which are generated by the corresponding canonical GO's. Table 3 summarizes the relation between GO's and SO sets. It lists the number of SO's produced by each GO for each set of valence AO's. On the basis of this table we now make the following analysis of the possible bonding interactions.

Since the f and g GO's produce SO's only in the tangential fluorine set and not on the cobalt, SO_{13} to SO_{18} become non-bonding MO's. The s, d0 and d2′ GO's produce SO's in the radial fluorine set and on cobalt but not in the tangential set. Interactions between the former two sets yield a BMO and an ABMO for each of these three cases due to "sigma-type" bonding. The d1′, d1″ and d2″ GO's produce SO's in the tangential fluorine set and on cobalt but not in the radial set. Interaction between the former

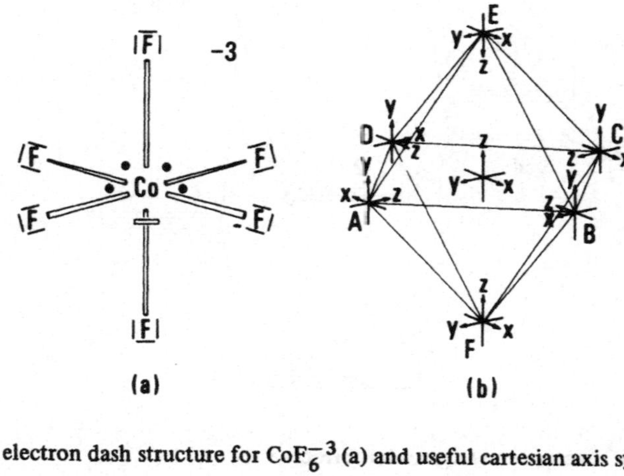

Fig. 24. An electron dash structure for CoF_6^{-3} (a) and useful cartesian axis systems for this ion (b).

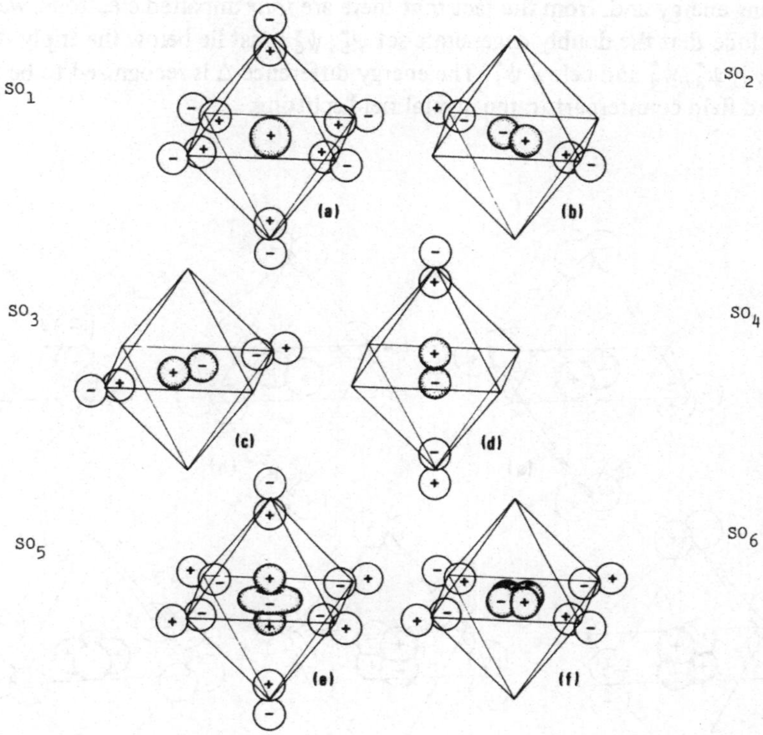

Fig. 25. Sigma-type symmetry orbitals for the CoF_6^{-3} ion obtained by means of generator orbitals.

two sets in this case will also yield bonding and antibonding MO's. However, these are "pi-type" interactions which yield smaller level splittings than the "sigma-type" interactions. For simplicity we shall ignore the "pi-type" splitting, which implies that the tangential fluorine AO's and the cobalt AO's do not mix. Each of the p0, p1′ and p1″ GO's produce SO's in all three SO sets. It is to be expected that interactions among them lead to a bonding, a non-bonding and an antibonding MO in each case. Again assuming that pi interactions are weaker than sigma interactions, we expect the non-bonding MO's to have their largest contribution from the tangential set. As before we therefore neglect any mixing between these tangential SO's and the radial SO's and the cobalt SO's. Thus all the tangential SO's are non-bonding MO's in this presentation. The same holds for the cobalt SO's generated by d1′, d1″ and d2″ and the only bonding interactions we consider are of the sigma type which are induced by the GO's s, d0, d2′, p0, p1′ and p1″, between the cobalt and radial fluorine orbitals (see Table 3).

Taking into account all of the foregoing considerations, we arrive at the qualitative energy level diagram in Fig. 27. From the experimental observation that this ion is paramagnetic, it can be concluded that Δ must be small compared with the electron pairing energy and, from the fact that there are *four* unpaired electrons, we further conclude that the doubly degenerate set ψ_5^*, ψ_6^* must lie below the triply degenerate set ψ_2^*, ψ_3^*, ψ_4^* and below ψ_1^*. The energy difference Δ is recognized to be the ligand field counterpart to the crystal field splitting.

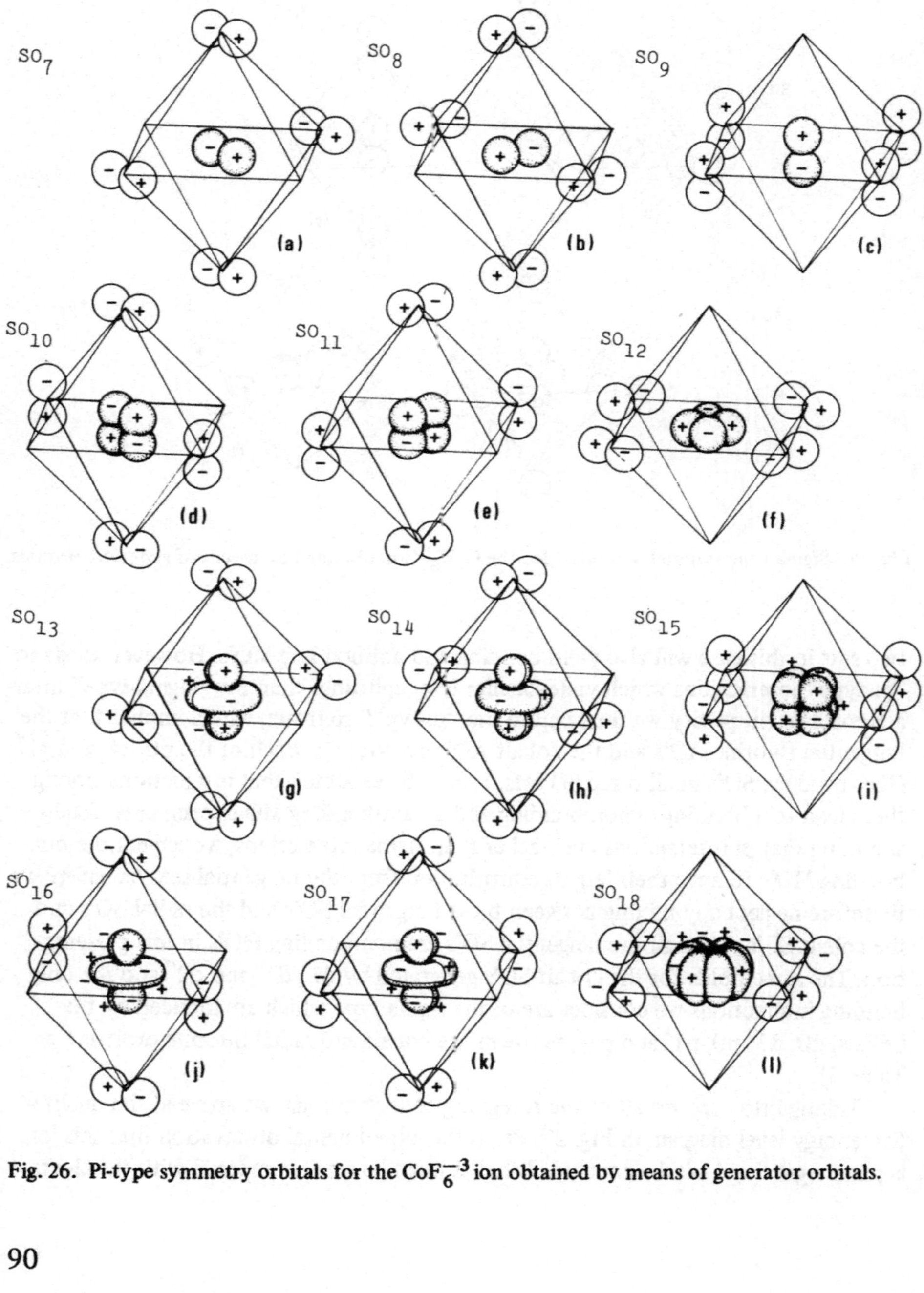

Fig. 26. Pi-type symmetry orbitals for the CoF_6^{-3} ion obtained by means of generator orbitals.

Table 3. The number of SO's generated by GO's in each set of CoF_6^{-3} valence AO's[a])

Symmetry Orbitals	Generator Orbitals										
	s	d0	d2′	d1′	d1″	d2″	p0	p1′	p1″	f′s	g′s
Valence AO's on Co	1	1	1	1	1	1	1	1	1	0	0
Radial set (Fpz)	1	1	1	0	0	0	1	1	1	0	0
Tangential set (Fpx, Fpy)	0	0	0	1	1	1	1	1	1	3	3
Equivalent irreducible representation	A_{1g}	E_g		T_{1g}			T_{2u}			T_{1u}	T_{2g}

[a]) The GO columns containing the same numbers are grouped together. Those with a knowledge of group theory will recognize that the left-to-right order of the columns so grouped corresponds to the irreducible representations shown. The dotted line separating the tangential SO's from the radial fluorine SO's and the cobalt SO's indicates the neglect of "pi-type" bonding interactions in the present discussion.

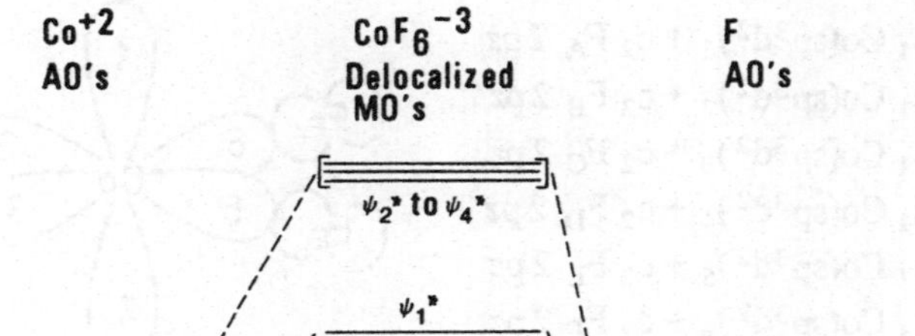

Fig. 27. A delocalized MO energy level diagram for CoF_6^{-3}.

11. Localized MO's of CoF_6^{-3}

In formulating a localized approach, we note that an s, three p and five d canonical GO's are associated with the occupied levels in the delocalized MO scheme of Fig. 27 and Table 3. The Lewis structure in Fig. 24 (a) suggests that there will be a set of six localized bonding MO's, one lone pair and four lone singles.

We consider first the six lowest delocalized *bonding* MO's ψ_1 to ψ_6 which were generated by the s, the three p, the d0 and the d2′ GO's. These six GO's can be hybridized to an octahedrally oriented sp^3d^2 set which is used to produce the localized MO's from the delocalized MO's. Taking immediately the crude but useful intuitive two-center localized approximation, we note that the hybrid GO's will generate an almost identical set of cobalt valence AO's and a 2*pz* AO on each fluorine. These can then be combined to form the following six localized MO's:

$$\phi_1 = c_1\,Co(sp^3d^2)_1 + c_2\,F_A\,2pz$$
$$\phi_2 = c_1\,Co(sp^3d^2)_2 + c_2\,F_B\,2pz$$
$$\phi_3 = c_1\,Co(sp^3d^2)_3 + c_2\,F_C\,2pz$$
$$\phi_4 = c_1\,Co(sp^3d^2)_4 + c_2\,F_D\,2pz$$
$$\phi_5 = c_1\,Co(sp^3d^2)_5 + c_2\,F_E\,2pz$$
$$\phi_6 = c_1\,Co(sp^3d^2)_6 + c_2\,F_F\,2pz$$

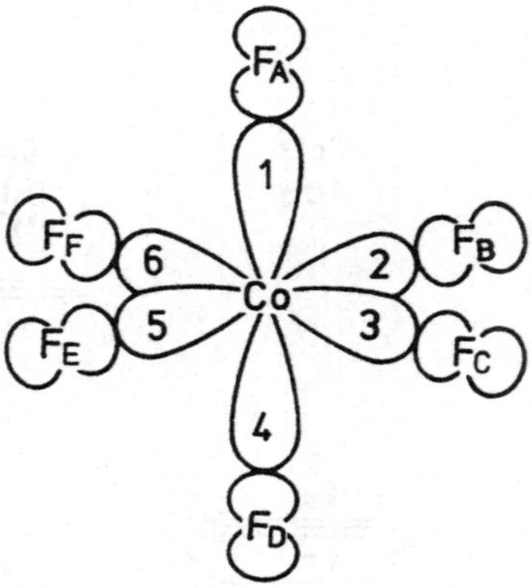

Localizing the tangential fluorine set regenerates essentially the 2 *px* and 2 *py* AO's on each fluorine, each one being filled by a lone pair.

With the remaining highest five occupied delocalized MO's stemming from the GO's d0, d2′ and d1′, d1″, d2″, we now wish to make four equivalent localized MO's to house the lone singles and one localized MO for the lone pair. Although any of the three AO's d1′, d1″ and d2″ could be used for the lone pair, the d2″ is most convenient because the mixing of the remaining d GO's is easy to visualize. Fig. 28 shows the mixing procedure which results in four equivalent d^4 GO's. We use these GO's to generate four equivalent localized MO's from the delocalized MO's d1′, d1″, ψ_5^*, ψ_6^* of Figure 27.[7]). The antibonding character, arising from the fluorine contribu-

[7]) Although these four hybrids are equivalent (which is a desired goal in localizing like electrons or electron pairs) they are only slightly more localized than the AO's from which they are constructed.

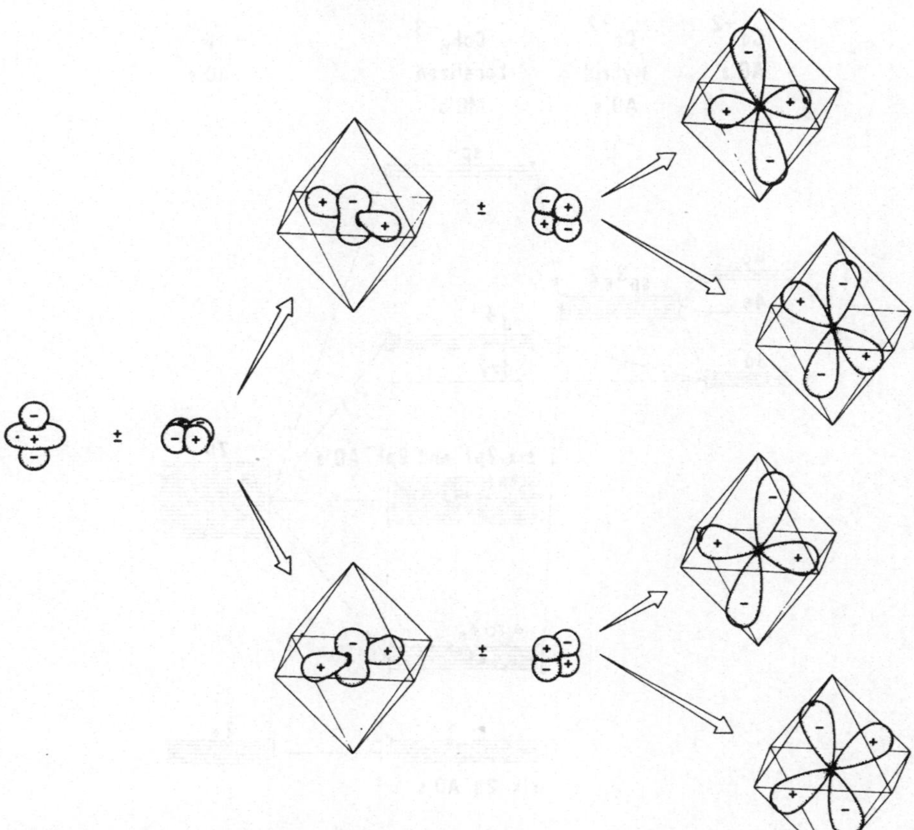

Fig. 28. Stepwise formation of a system of four equivalent d AO's.

tions in ψ_5^* and ψ_6^*, will be evenly distributed over these four equivalent localized MO's showing that some of the negative charge resides on the fluorines. Since these levels are energetically close to the cobalt 3 d levels, we expect these orbitals to be largely localized on cobalt.

Finally let us also localize the unoccupied delocalized MO's ψ_1^* to ψ_4^* of Fig. 27 and Table 3. Since they are generated from an s and three p GO's, the appropriate hybrid GO's clearly are sp^3. The resulting localized MO's are most easily visualized as cobalt sp^3 hybrids directed towards four of the eight octahedral faces and having antibonding interactions with the three fluorines on the corners of that face. In Fig. 29 is shown an energy level diagram in which the above considerations are reflected.

A very satisfying feature of the described localized view of transition metal complexes such as CoF_6^{-3} is that there is no necessity to invoke cobalt 4 d orbitals in order to preserve the four unpaired electrons, as is done in the familiar Pauling-type valence bond approach (*8*).

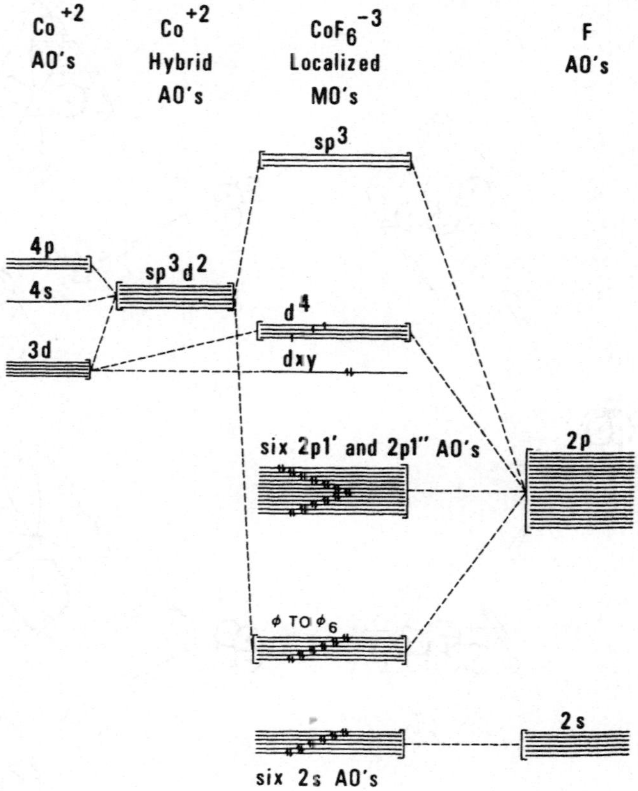

Fig. 29. A localized MO energy level diagram for CoF_6^{-3} excluding pi interactions.

12. Theoretical Considerations

By means of these selected examples, we have attempted to show the simplicity and wide applicability of the generator orbital approach. It provides a useful delocalized bonding view which has a degree of rigor comparable to that provided by the more formal methods of group theory; it also yields an equivalent localized description (*9*).

In order to recognize that the GO approach indeed accomplishes the desired group theoretical goal, we note that the *atomic* invariance group is the full three-dimensional rotation-reflection group O_3, and that this group contains all *molecular* point groups as subgroups (*10*). Since the canonical atomic orbitals (*nlm*) form bases for the irreducible representations of O_3, they automatically also form bases for representations of the molecular point groups. It should be emphasized that this property depends only upon the *angular* parts of the atomic orbitals, namely, the spherical harmonics $Y_{lm}(\theta\Phi)$. Therefore, only the AO's with lowest quantum number n, i.e., orbitals without radial nodes, are needed for the GO method.

With regard to the localized orbitals, we note that a many-electron MO wave function is invariant under any orthogonal transformation among MO's with the same occupation number (*11*). Therefore the determination of localized MO's requires finding appropriate orthogonal transformations. We deduce these from hybrid GO expansions in terms of canonical GO's. The justification for this procedure is that in systems of high symmetry, localized MO's are known to be identical with the so-called equivalent MO's (*12*).

Finally it is appropriate to comment on the relationship between GO's and united atom orbitals (*13*). A canonical GO can be considered to be the angular part of the corresponding united atom orbital. However, the properties of GO's are of general applicability whereas the united atom orbitals are actually poor models for valence MO's. This is so because the GO concept represents a simple embodiment of the representational properties of the molecular point group, whereas the united atom concept contains additional detail resulting from a relatively poor physical model designed to simplify the description of the interactions between nuclei and electrons. Consequently united atom orbitals have very specific radial nodal structures. These nodes are unrealistic, are irrelevant for our purposes, and can only be deduced once the united atom MO approximation is known. By contrast we have shown that canonical GO's can be used to construct canonical MO's.

It is apparent that the methods discussed here are particularly effective for molecules whose MO's are largely determined by their point group symmetry because, in these cases, central AO's are good MO templates. Clearly, there are other types of molecules in which central AO's are less suitable and other simple orbitals are more effective templates. For general pi-electron systems, for example, free-electron MO's would be more useful GO's (*14*). In its broadest sense, then, the generator orbital approach is based on the principle that major characteristics of complicated molecular orbitals can be semi-quantitatively anticipated by conceptual analogy to simpler orbitals with which they share essential physical features. We feel that there is an elucidating quality in this reduction of the more complex to the less complex, which is of value to theorists and non-theorists alike.

Acknowledgement. This work was supported by the U.S. Energy Research and Development Administration, Division of Physical Research and the Department of Chemistry.

References

1. *Richards, W.G., Walker, T.E.H., Hinkley, R.K.:* A Bibliography of ab initio Molecular Wave Functions. Oxford: Clarendon Press, 1971.
2. *Richards, W.G., Walker, T.E.H., Farnell, L., Scott, P.R.:* A Bibliography of ab initio Molecular Wave Functions: Supplement for 1973. Oxford: Clarendon Press, 1974.
3. See for example, *Cotton, F.A.:* Chemical Applications of Group Theory, 2nd Ed. New York: John Wiley & Sons, Inc., 1971.
4. *England, W., Salmon, L.S., Ruedenberg, K.:* Topics in Current Chemistry, 23, 31 (1971).
5. *Pimentel, G.C., Spratley, R.D.:* Chemical Bonding Clarified Through Quantum Mechanics. San Francisco: Holden-Day Inc., 1969.
6. *Gray, H.B.:* Chemical Bonds. California: Benjamin, W.A., Menlo Park, 1973.
7. *Hoffman, D.K., Ruedenberg, K., Verkade, J.G.:* J. Chem. Educ., in press.
8. *Pauling, L.:* The Nature of the Chemical Bond, 3rd ed., Ithaca, NY: Cornell University Press, 1960. A comparison of the valence bond approach with other theories as applied to transition metal complexes can be found in *Huheey, J.E.:* Inorganic Chemistry: Principles of Structure and Reactivity. New York: Harper and Row, 1972.
9. The ideas outlined here are elaborated in a forthcoming text by the authors to be published by Springer-Verlag.
10. a) *Tinkham, M.:* Group Theory and Quantum Mechanics. McGraw-Hill, 1964;
    b) *Hamermesh, M.:* Group Theory and its Application to Physical Problems. Addison Wesley, 1962.
11. *Fock, V.:* Z. Physik, 61, 126 (1930).
12. *Lennard-Jones, J.E.:* Proc. Roy. Soc. (London) *A 198*, 1, 14 (1949). However, the *most* localized MO's do not necessarily form an equivalent set (see *Edmiston, C., Ruedenberg, K.:* In: Quantum Theory of Atoms, Molecules and the Solid State, (Lowdin, P.O., Ed.). Academic Press, 1966, p. 263).
13. The united atom concept goes back to R.S. Mulliken, in connection with his invention of the correlation diagram (*Mulliken, R.S.:* Rev. Mod. Phys., 2, 60, 506 (1930) and 4, 40 (1932)). Mulliken did not suggest the united atom model as a reliable approximation near the equilibrium conformation, however; a reservation borne out by later calculations.
14. *Platt, J.R., Ruedenberg, K., Scherr, C.W., Ham, N.S., Labhart, H., Lichten, W.:* Free-Electron Theory of Conjugated Molecules: A Source Book. Wiley, J. & Sons, Inc., 1964.

Ligand Field Theory of f-Orbital Sandwich Complexes

Keith D. Warren

Department of Chemistry, University College, Cardiff, Wales, U.K.

Table of Contents

1. Introduction

Although the elements of the lanthanide and actinide series have long been known to exhibit a quite extensive organometallic chemistry, it is only within the last decade that typical sandwich species have been prepared and studied. These systems however, although resembling the familiar metallocene and bis-arene compounds of the d-block elements, are not strictly their analogues since in both f-orbital series the known sandwich complexes are derived only from the cyclooctatetraenyl dianion. This survey is therefore concerned with the properties of the $M^{IV}(Cot)_2$ species of the 5f series and the $[M^{III}(Cot)_2]^-$ anions of the 4f and 5f series, wherein the microsymmetry about the metal atom is either D_{8h} or D_{8d} (vide infra).

The original discovery of these systems was not however entirely accidental but arose following a persistent theoretical probing into the question of the part played by f-orbitals in chemical bonding. Thus, after the early hybridisation treatment, due to *Coulson* and *Lester* (*1*), the problem was attacked by *Moffitt* (*2*) on a molecular orbital basis, and more recently by *Kettle* and *Pioli* (*3*), using an approach remeniscent of the angular overlap model of *Jørgensen*, *Pappalardo*, and *Schmidtke* (*4*), and by *Kettle* and *Smith* (*5*), via a group theoretical treatment of the behaviour of the f-orbital set in various molecular geometries (c.f. also *Eisenstein* (*6*)).

Nevertheless, the existence of biscyclooctatetraenyl metal complexes, bonded via the metal f-orbitals and the aromatic π-systems of the rings, was first predicted on the basis of Wolfsberg-Helmholz MO calculations by *Fischer* (*7*) in 1963, who treated the then hypothetical $U(Cot)_2$ molecule, and demonstrated experimentally by *Streitwieser* and *Müller-Westerhoff* (*8*) in 1968. These authors succeeded in synthesising the uranium derivative, $U(Cot)_2$, and gave to it the trivial name, uranocene, in order to underline its resemblance to the d-series metallocenes both structurally and, as they proposed, in its mode of metal-ring bonding.

Since 1968 extensive studies of the 5f analogues of $U(Cot)_2$ have been reported (*9–15*), and $An^{IV}(Cot)_2$ complexes are now also known for Th, Pa, Np, and Pu, together with a number of alkyl- and aryl-substituted derivatives (*16–19*). Moreover, biscyclooctatetraenyl derivatives have also been obtained for several of the lanthanide elements and $[Ln^{III}(Cot)_2]^-$ species are now known for La, Ce, Pr, Nd, Sm, Gd, and Tb (*20, 21*). For $U(Cot)_2$ the sandwich structure originally proposed (*8*) has been confirmed by a crystallographic determination (*22*), which showed the complex to possess D_{8h} symmetry. Similar measurements on the potassium salt of the $[Ce(Cot)_2]^-$ anion also revealed an eight-fold rotation axis of symmetry, but in this case the staggered D_{8d} conformation was found (*23*), rather than the eclipsed D_{8h} structure. It

---

Abbreviations used:

Cot = C_8H_8
EtCot = $C_2H_5 \cdot C_8H_7$
n-BuCot = n-$C_4H_9 \cdot C_8H_7$
C_6H_5Cot = $C_6H_5 \cdot C_8H_7$
Me_4Cot = 1,3,5,7-$(CH_3)_4 \cdot C_8H_4$
$(C_6H_5)_4$Cot = 1,3,5,7-$(C_6H_5)_4 \cdot C_8H_4$
Cp = C_5H_5.

seems probable that the potential barrier between these two conformations is very small (as in the metallocenes) since the degree of metal-ligand mixing is not significantly affected, and for simplicity the nomenclature of the D_{8h} point group will be generally employed for the biscyclooctatetraene complexes discussed. Further details of the structural chemistry of the cyclooctatetraene complexes of the 4f and 5f series are however given in the excellent review of *Baker*, *Halstead*, and *Raymond* (*24*).

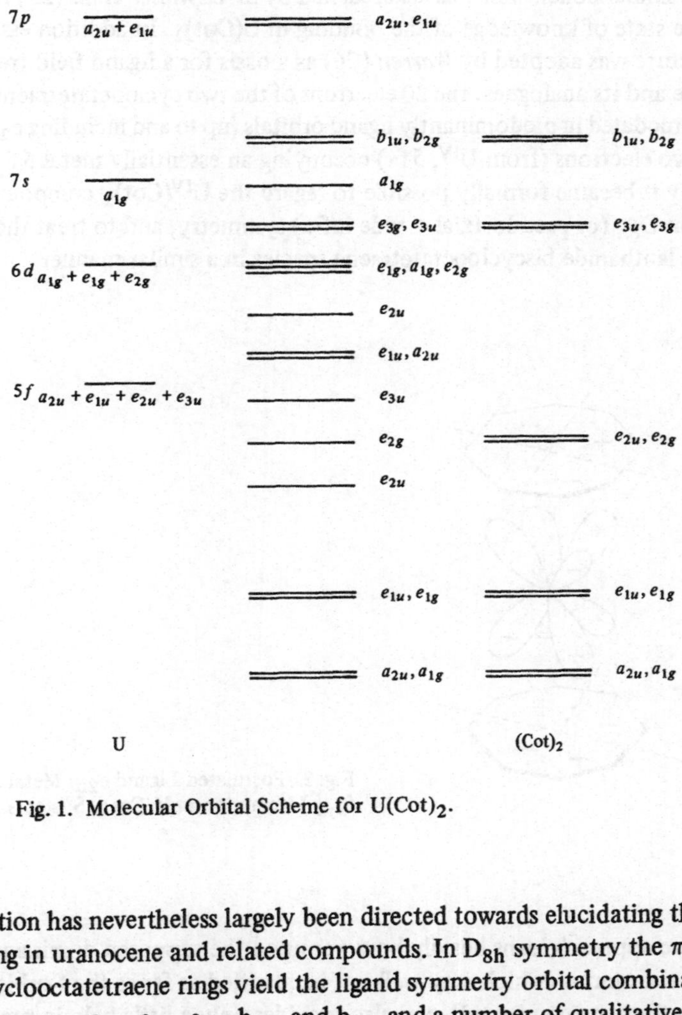

Fig. 1. Molecular Orbital Scheme for $U(Cot)_2$.

Attention has nevertheless largely been directed towards elucidating the nature of the bonding in uranocene and related compounds. In D_{8h} symmetry the π-orbitals of the two cyclooctatetraene rings yield the ligand symmetry orbital combinations a_{1g}, a_{2u}, e_{1u}, e_{1g}, e_{2u}, e_{2g}, e_{3u}, e_{3g}, b_{1u}, and b_{2g}, and a number of qualitative or semi-quantitative molecular orbital schemes have used the energy ordering deduced by *Fischer* (*7*) as a starting point, i.e.

$$a_{2u} \sim a_{1g} < e_{1u} \sim e_{1g} < e_{2u} \sim e_{2g} < e_{3u} \sim e_{3g} < b_{1u} \sim b_{2g} \quad \text{(see Fig. 1).}$$

For the metal orbitals the energetic order will clearly be 5 f $<$ 6 d $<$ 7 s $<$ 7 p, and it has been assumed (*25, 26*) that only the lowest of these – the 5 f orbitals – will be appreciably involved in the bonding. In carrying out molecular orbital calculations however the main difficulty has been found in deciding upon an appropriate one-electron core term (H^{core}) for the metal f-orbitals, relative to the H^{core} terms given by *Fischer* (*7*) for the ligand symmetry orbitals. Thus, in order to avoid the accumulation of negative charge on the metal atom it was assumed by *Hayes* and *Edelstein* (*27*) that the metal f-level lay between the ligand e_2 and e_3 pairs of levels, as shown in Fig. 1, and a similar conclusion was also reached by *Streitwieser et al.* (*25*) in their survey of the state of knowledge of the bonding in $U(Cot)_2$. In addition essentially the same picture was adopted by *Warren* (*26*) as a basis for a ligand field treatment of uranocene and its analogues, the 20 electrons of the two cyclooctatetraenyl dianions being accommodated in predominantly ligand orbitals (up to and including e_{2g}) with the remaining two electrons (from U^{IV}, $5f^2$) occupying an essentially metal 5 f orbital. Consequently it became formally possible to regard the $U^{IV}(Cot)_2$ compound as a $5f^2$ system in D_{8h} (or pseudo-axial – vide infra) symmetry, and to treat the other actinide and lanthanide biscyclooctatetraene species in a similar manner.

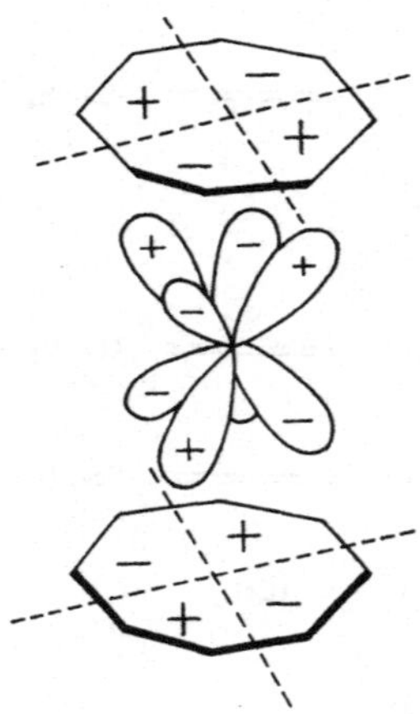

Fig. 2. Postulated Ligand e_{2u}: Metal $f_{z(x^2-y^2)}$, f_{xyz} Interaction in $M(Cot)_2$ Systems.

The above approach leans heavily however upon the supposed dominance of the one-electron core terms (which may well not be justified as far as the f-orbital – ligand interactions are concerned), and also provides only a little help in predicting the energetic ordering of the f-levels in D_{8h} symmetry. Clearly if the one-electron picture is assumed with

$$e_{2u}(\text{ligand}) < 5f(\text{metal}) < e_{3u}(\text{ligand}),$$

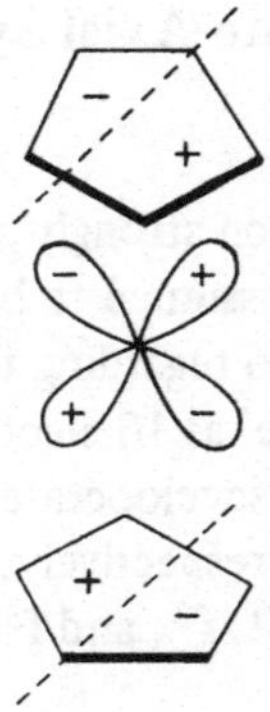

Fig. 3. Ligand e_{1g}: Metal d_{xz}, d_{yz} Interaction in $M(Cp)_2$ Systems.

the mainly metal 5f, e_{3u}, level should be bonding in character whilst the remaining 5f levels (a_{2u}, e_{1u}, and e_{2u}) should be anti-bonding, and *Streitwieser et al.* (*8, 25*) have argued strongly that the e_{2u} metal 5f – ligand interaction is the most significant contribution to the bonding in such species, by analogy with the known importance of the e_{1g} metal 3d – ligand interaction in metallocene systems (see Figs. 2 and 3). This latter suggestion is supported by the fairly large group overlap integrals found for the e_{2u} level (vide infra), and on this basis therefore the e_{2u}, mainly metal 5f, level should be the most markedly anti-bonding in character.

A further caveat to the simple one-electron picture described above also emerges from INDO SCF calculations on the $[Ce(Cot)_2]^-$ system carried out by *Clack* and *Warren* (*28*). This complex has formally a $4f^1$ configuration so that no f : f repulsion terms are involved, but it was nevertheless found that the total energy of the system was influenced to a marked extent by the degree of mixing of the metal 5d and the ligand orbitals, and the d : f repulsion terms arising therefrom. Since the overall splitting of the f-orbitals is likely only to amount to a few hundred cm^{-1} in the 4f series (*26*) it is clear that such two-electron terms may well have as significant an effect in determining total energies as do the one-electron terms depending upon H^{core} quantities and overlap integrals. Thus, at least in the 4f series, an estimate of the f-orbital splitting based only on H^{core} terms and overlaps may be somewhat suspect, although for the 5f series, where the overall splitting is likely to be rather larger (some 2 to 4 kK) it should be more reliable. The problems attending the estimation of the splitting parameters for the f-orbital set will however be discussed further in Section 2, in which the necessary ligand field formalism is developed.

2. Ligand Field Theory of f^x Configurations in Pseudo-Axial Symmetry

The molecular orbital approach outlined in the previous Section strongly suggests that a ligand field treatment of the highest occupied levels – assumed to be mainly f-orbitals – might be fruitful, and in effect this is equivalent to regarding the two cyclooctatetraene rings of the $An(Cot)_2$ species as $(Cot)^{2-}$ (i.e. as 10 electron systems) with the metal formally An^{IV}. The Pa, U, Np, and Pu biscyclooctatetraene derivatives thus correspond to f^1, f^2, f^3, and f^4 configurations respectively, whilst for the lanthanides the $[Ln^{III}(Cot)_2]^-$ anions similarly represent f^1, f^2, and f^3 configurations for the Ce, Pr, and Nd systems.

As noted in Section 1 both $An^{IV}(Cot)_2$ and $[Ln^{III}(Cot)_2]^-$ species are known to possess eight-fold rotational symmetry and it is therefore profitable to generalise the treatment of d^x configurations in various point groups due to *Scott* and *Matsen* (*29*). In considering d-block sandwich complexes these authors showed that d^x configurations could be handled in terms of a field of effectively axial, $C_{\infty v}$, symmetry for the groups C_n, C_{nh}, C_{nv}, D_n, D_{nh}, and D_{nd} as long as n was greater than or equal to 5, and it is readily found that for any l^x configuration a similar result follows for $n \geqslant 2l + 1$. Thus for f^x configurations pseudo-axial symmetry may be assumed for $n \geqslant 7$, as in the point groups D_{8h} and D_{8d}.

The situation differs markedly however from that of the d^x configurations of metallocenes and bis-arene systems for which pseudo-axial symmetry also obtains. In these latter cases the pseudo-axial field is substantial and generally will dominate the interelectronic repulsion terms which in turn are usually more significant than the spin-orbit interactions, i.e. $V_{LF} > e^2/r_{12} > \xi\, l.s$. In the 4f series however it is normally found that the ligand field effects are very small because of the inner, highly shielded, spatial disposition of the f-orbitals, and consequently the electron repulsion and spin-orbit terms predominate, i.e. $e^2/r_{12} > \xi\, l.s > V_{LF}$. In the 5f series, because of the more radially extended orbitals, ligand field splittings become more important, but the spin-orbit coupling constants are also larger and the order of effects is roughly $e^2/r_{12} \sim \xi\, l.s > V_{LF}$.

Consequently, in both the 4f and 5f series, the spin-orbit and interelectronic repulsion terms appreciably exceed the ligand field contributions and for both f^x series the use of a weak field basis set with explicit inclusion of spin-orbit effects is therefore necessary. The functions needed are thus the $|LSJM_J\rangle$ set, obtained via the Clebsch-Gordan coefficients from the $|LSM_LM_S\rangle$ set, and for the 4f series J is generally a good quantum number, with ground states of high purity in the $|LSJM_J\rangle$ representation. For the 5f series the larger ligand field leads to greater J mixing, and in any case the stronger intermediate coupling leads to lower purity ground states, ranging from some 95% for $5f^2$ to only 81% for $5f^4$.

In the absence of a magnetic field the appropriate perturbation Hamiltonian, H′, is given by

$$H' = \sum_{i<j} \frac{e^2}{r_{ij}} + \sum_i \xi(r_i)\, l_i \cdot s_i + \sum_i V_{LF}(r_i)$$

where the three terms represent respectively the interelectronic repulsions, the spin-orbit contributions, and the effective ligand field. In pseudo-axial symmetry this latter may be written as

$$V_{LF} = c_1 Y_0^0 + c_2 Y_2^0 + c_3 Y_4^0 + c_4 Y_6^0$$

where the Y_k^q are spherical harmonics of order q and degree k, and the c_n quantities weighting coefficients. The Y_0^0 term will not split the f-orbital set (and may therefore be disregarded), and since q = 0 throughout V_{LF} will not mix orbitals of differing m_l values and the V_{ax} contributions will be diagonal in m_l. Thus, in $C_{\infty v}$ symmetry, the f-orbital set yields four irreducible representations, Σ^-, Π, Δ, and Φ, corresponding to $m_l = 0, \pm 1, \pm 2$, and ± 3 respectively, and one may define the one-electron ligand field parameters

$$\langle 0|V_{ax}|0\rangle = \sigma, \quad \langle \pm 1|V_{ax}|\pm 1\rangle = \pi, \quad \langle \pm 2|V_{ax}|\pm 2\rangle = \delta,$$

and

$$\langle \pm 3|V_{ax}|\pm 3\rangle = \varphi.$$

When spin-orbit coupling is explicitly included the $|LSJM_J\rangle$ wave functions are characterised according to the irreducible representations of the double group, C_∞^*, as defined by the relevant M_J values. Thus $M_J = 0$ corresponds to the representations Σ^+ or Σ^- (J even or odd), and $M_J = \pm 1, \pm 2, \pm 3, \ldots$ to the representations Π, Δ, $\Phi, \ldots$. For half-integral M_J values, $\pm 1/2, \pm 3/2, \pm 5/2, \ldots$ the correspondence is to the representations $\Pi^*, \Delta^*, \Phi^*, \ldots$ as previously defined (*30*).

It is not however to be expected that the behaviour of the various f^x systems in pseudo-axial symmetry should be dominated by the one-electron energy differences; indeed the states arising will certainly not be characterised by well defined orbital occupations but will, especially in the lanthanide series, constitute a situation more closely analogous to that obtaining for systems with large internuclear distances (see, for example, *Jørgensen* (*31*)), in which substantial off-diagonal matrix elements of e^2/r_{12} or $\xi\, l.s$ will overwhelm diagonal differences in the V_{LF} terms. Consequently the weak field treatment adopted is the only feasible approach for lanthanide systems, and even for the actinides leads to states (especially ground levels) of much greater purity than the strong field scheme.

Nevertheless, although the V_{LF} f-orbital splitting cannot be such a dominant effect, it is desirable to examine more closely the means which may be adopted to predict both the energetic ordering of the f-orbitals and the magnitude of the resultant splitting. Consideration is now therefore given first to the classical electrostatic crystal field model and thereafter to methods based on molecular orbital calculations and related schemes. As will be seen the former approach proves to be phenomenologically useful but quantitatively inadequate, whilst, because of the homomorphous relationship between the crystal field model and the molecular orbital approach, the latter may ultimately prove the more valuable. (Compare, for example, the use of the angular overlap model to treat various f-orbital systems in terms of one-electron orbital energy differences (*4*).)

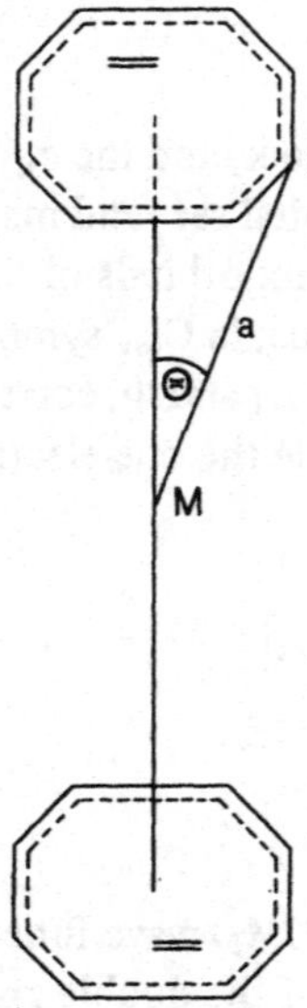

Fig. 4. Crystal Field Model for $M(Cot)_2$ Systems.

For the crystal field model one assumes (see Fig. 4) that each carbon atom of the two cyclooctatetraenyl dianion rings will bear a negative charge, z_i, of 1/4 with

$$V_{ax} = \sum_{i=1}^{16} z_i e^2/r_{ij} .$$

Thus expansion in spherical harmonics about the origin yields

$$(r_{ij})^{-1} = (\pi/5)^{1/2} (r^2/a^3) (3 \cos^2\theta_i - 1) Y_2^0 + (\pi/9)^{1/2} (r^4/a^5) (35 \cos^4\theta_i - 30 \cos^2\theta_i + 3) Y_4^0 + (\pi/13)^{1/2} (r^6/a^7) (231 \cos^6\theta_i - 315 \cos^4\theta_i + 105 \cos^2\theta_i - 5) Y_6^0 ,$$

and defining the parameters

$$Ds = (8/15) (\langle r^2 \rangle/a^3) (3 \cos^2\theta - 1) z e^2 ,$$

$$Dt = (2/33) (\langle r^4 \rangle/a^5) (35 \cos^4\theta - 30 \cos^2\theta + 3) z e^2 ,$$

and

$$Du = (5/429) (\langle r^6 \rangle/a^7) (231 \cos^6\theta - 315 \cos^4\theta + 105 \cos^2\theta - 5) z e^2 ,$$

one finds (*26*)

$$\langle\ 0|V_{ax}|\ \ 0\rangle^{*} = 4\,Ds + 6\,Dt + 20\,Du$$
$$\langle\pm 1|V_{ax}|\pm 1\rangle = 3\,Ds + Dt - 15\,Du$$
$$\langle\pm 2|V_{ax}|\pm 2\rangle = -7\,Dt + 6\,Du$$

and

$$\langle\pm 3|V_{ax}|\pm 3\rangle = -5\,Ds + 3\,Dt - Du\,.$$

Using the operator equivalent technique of *Stevens* (*32*), and following *Elliott* and *Stevens* (*33*), one may also express the splittings in terms of the better known $A_k^q\langle r^k\rangle$ parameters (c.f. also *Abragam* and *Bleaney* (*34*)), which are in turn simply related to the B_q^k parameters which occur in the tensor operator treatment of crystal field splittings. Using the relationship given by *Wybourne* (*35*), the crystal field and B_q^k parameters are found to be connected by the expressions

$$B_0^2 = 15\,Ds\,,\quad B_0^4 = 33\,Dt\,,$$

and

$$B_0^6 = (429/5)\,Du\,.$$

It should however be noted that for the lanthanide and actinide series numerous different definitions of the crystal field parameters have on occasion been adopted and care should therefore be taken in comparing quantities from different literature sources. In this review the crystal field parameters are throughout used in the sense defined above, but see for example *Hutchings* (*36*) for a survey of some other conventions.

For lanthanide systems, as for the d-transition series, a purely electrostatic calculation is not expected to yield realistic values for the B_q^k parameters. Thus for $PrCl_3$ *Hutchings* and *Ray* (*37*) found that only B_0^4 of the calculated quantities was in reasonable agreement with the experimentally determined results, the calculated B_0^2 and B_0^6 values being respectively much larger and much smaller than the experimental results. Moreover, for $[Ce(Cot)_2]^-$, using the crystallographic results (*23*) a = 2.742 Å and θ = 40.85°, and the $Ce^{3+}\langle r^k\rangle$ values of *Watson* and *Freeman* (*38*), and for $U(Cot)_2$ with a = 2.647 Å and θ = 43.40° (*22*) and the $\langle r^k\rangle$ of *Lenander* (*39*), similar conclusions follow (see (*26*)).

Consequently, although the crystal field model is of great utility in representing phenomenologically the f-orbital splitting pattern, it is quite inadequate for any attempt to calculate the Ds, Dt, and Du (or B_0^2, B_0^4, and B_0^6) parameters, and a more promising approach is to consider the admittedly small f-orbital – ligand interactions and to determine how these may be distinguished according to the irreducible representations to which the f-orbitals belong. In practice therefore a molecular orbital treatment is required but apart from the preliminary SCF calculation of *Clack* and *Warren* (*28*) only Wolfsberg-Helmholz type treatments have been reported for systems of pseudo-axial symmetry (*15, 26, 27*), whose results will necessarily be somewhat questionable for systems in which the metal-ligand interactions are weak. Nevertheless, lacking more sophisticated calculations there is no alternative but to accept, cum grano salis, the results of the one-electron calculations: rather surprisingly the magnetic measurements appear to suggest that the orbital scheme thus deduced may not be too improbable.

For a one-electron calculation aimed at estimating the f-orbital splitting only a limited basis set need be employed – the f-orbitals of the metal and the π-type carbon 2p orbitals of the conjugated ring systems. In a detailed treatment the other orbitals (2s and 2p) of the ring carbon atoms should be included since, as pointed out by *Hillier* and *Canadine* (*40*) the traditional σ–π separability conditions are not satisfied in sandwich complexes, but with the inherent approximations of the Wolfsberg-Helmholz method such a refinement would be superfluous. Using the nomenclature of the D_{8h} point group the ligand π-orbitals a_{2u}, e_{1u}, e_{2u}, and e_{3u} (see Fig. 1) may severally interact with the corresponding f-orbitals of the same symmetry, and the f-orbital splitting will therefore depend on the relative one-electron energies of the metal f-orbitals and the ligand π-orbital symmetry combinations, and upon the corresponding group overlap integrals. Calculations of this type have in fact been made by three sets of authors – by *Hayes* and *Edelstein* (*27*) and by *Amberger, Fischer*, and *Kanellakopulos* (*15*) using the basic Wolfsberg-Helmholz technique, and by *Warren* (*26*) using *Jørgensen's* (*31*) variant in which the metal and ligand orbitals are first orthogonalised by the Schmidt method, so as to make second-order perturbation procedures directly applicable – and the results of these treatments will be considered below.

For many metal complex systems an alternative approach via the group overlap integrals is possible using the Angular Overlap Model (*4, 41, 42*). Thus the Wolfsberg-Helmholz treatment, with k = 2 (i.e. $H_{ML} \doteq S_{ML}(H_M + H_L)$), allows the anti-bonding energies, E^*, of the central metal orbitals to be written as (*42*)

$$E^* = \frac{H_L^2}{(H_M - H_L)} \cdot G_{ML}^2$$

where G_{ML} is the appropriate group overlap integral between the metal orbital and the ligand symmetry orbital combination of the same irreducible representation. In many cases the overlap integrals may be resolved into a radial part, depending only on the metal and ligand radial functions and the bond length, and an angular part associated with the overlap of a metal orbital of azimuthal quantum number, l, with a ligand orbital of λ symmetry ($\lambda = \sigma, \pi, \ldots$) with respect to the metal-ligand bond, but unfortunately in metal sandwich somplexes the ligand π-orbitals are neither coincident with, nor at right angles to, the metal-carbon direction. Consequently, for a p-type ligand orbital the individual overlaps will in general contain both σ and π contributions, as will also the corresponding group overlaps, and a simple radial-angular resolution as outlined above is not applicable.

For this reason the use of the Angular Overlap Model made by *Amberger et al.* (*15*) seems inappropriate since E* was there taken to be proportional to the sum of the squares of the S_σ and S_π parts of G_{ML}, each multiplied respectively by a parameter, e_σ or e_π. When the σ and π contributions arise from *different* ligand orbitals their effects are naturally additive, but for sandwich species the individual S_σ and S_π components are due to overlap with the *same* ligand orbital, and in their overall effect these S terms may in general either reinforce or oppose each other. Thus the anti-bonding energy, E*, will depend upon the *sum* of the S_σ and S_π contributions to G_{ML} and be proportional only to G_{ML}^2. Moreover, the application of the Angular Overlap Model requires that the ligand symmetry orbitals of a given λ be at least approximately equienergetic, which is certainly not the case for sandwich species, and it is not indeed even clear a priori which metal orbitals will be bonding and which anti-bonding in character. Nevertheless, the evaluation of G_{ML} for f-orbital systems is crucial to any calculation, and the derivation is now therefore outlined, following the procedure of *Clack* and *Smith* (*43*), using the $f_{z^3}(a_{2u})$ metal orbital as an example.

For the calculation of $G_{ML}(a_{2u})$ it is first necessary to evaluate the general expression for the f_{z^3}–p_z overlap (see Fig. 5) in terms of the angles, *a* and *b*, and the two orbitals are therefore expressed in terms of component orbitals which would give rise to purely σ, π, ... overlaps along the line *AB*. This is achieved using the transform

$$\begin{bmatrix} x \\ y \\ z \end{bmatrix} = \begin{bmatrix} \cos a \cdot \cos b & -\sin a & \sin b \cdot \cos a \\ \cos b \cdot \sin a & \cos a & \sin b \cdot \sin a \\ -\sin b & 0 & \cos b \end{bmatrix} \begin{bmatrix} x' \\ y' \\ z' \end{bmatrix}$$

expressing the orbitals involved in Cartesian form. Thus the original molecular co-ordinates (x, y, z) are expressed in terms of a new set (x', y', z') whose *z*-axis lies along *AB*, with the *x*-axis in the plane *ABT*. The same transformation is then applied to the orbitals at both *A* and *B* and the pure σ, π, ... overlap components thus resulting evaluated. Note that the codirectionality of the primed coordinate sets at *A* and *B*

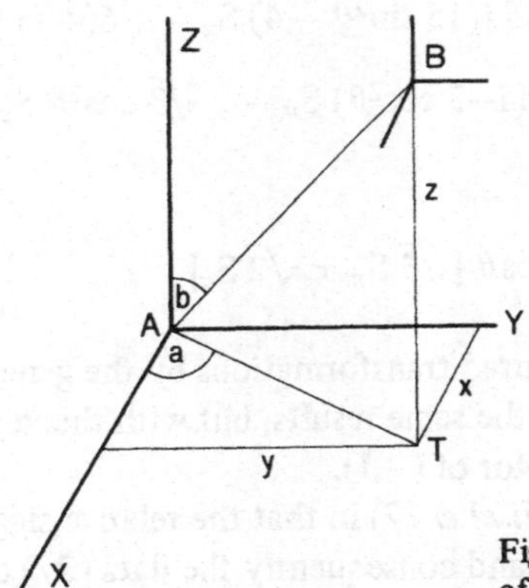

Fig. 5. Diagram illustrating Angular Dependence of Overlap.

differs from the standard *Mulliken* (*44*) convention for overlap integrals (in which the two z' axes point towards each other) and allowance for this must be made in the final results. Thus one finds, for a = 0, b = θ,

$$p_z \rightarrow -\sin b \cdot p'_x + \cos b \cdot p'_z$$

and

$$f_{z^3} \rightarrow \frac{1}{2} \cos b\,(5\cos^2 b - 3) \cdot f'_{z^3} - \frac{1}{4}\sqrt{6}\sin b\,(5\cos^2 b - 1).\ f'_{xz^2}$$

$$+ \frac{1}{2}\sqrt{15}\cos b \sin^2 b \cdot f'_{z(x^2-y^2)} - \frac{1}{4}\sqrt{10}\sin^3 b \cdot f'_{x(x^2-3y^2)}$$

but the δ and ϕ components of S clearly vanish leaving

$$S(p_z, f_{z^3}) = \frac{1}{2}\cos^2\theta(5\cos^2\theta - 3)\,\overline{S}_\sigma + \frac{1}{4}\sqrt{6}\sin^2\theta(5\cos^2\theta - 1)\,S_\pi$$

where $\overline{S}$ indicates that multiplication by the factor (– 1) is needed to conform to the Mulliken convention. Thus, using the ligand symmetry orbitals for the cyclooctatetraenyl dianion given by *Cotton* (*45*), and the relationships

$$\sum_{m=0}^{n-1} \cos^2 m(2\pi/n) = \sum_{m=0}^{n-1} \sin^2 m(2\pi/n) = n/2$$

where n is the number of atoms in each ring (here 8), one readily evaluates the contributions of the other individual orbitals ($b = \theta$, $a = m(2\pi/n)$), and obtains

$$G(a_{2u}),\ \sigma = \sqrt{6}\sin^2\theta(4\text{–}5\sin^2\theta)\,S_\pi - 2\cos^2\theta(2\text{–}5\sin^2\theta)\,S_\sigma,$$

where $S_\sigma = S(p_z, f_{z^3})$ and $S_\pi = S(p_x, f_{xz^2})$. Similarly for the other f-orbitals are obtained the results

$$G(e_{1u}),\ \pi = (1/\sqrt{2}) \cdot \sin\theta\cos\theta\,[(15\sin^2\theta - 4)\,S_\pi - \sqrt{6}(4\text{–}5\sin^2\theta)\,S_\sigma],$$

$$G(e_{2u}),\ = \frac{1}{2}\sqrt{10}\sin^2\theta\,[\sqrt{2}(1\text{–}3\cos^2\theta)\,S_\pi - 2\sqrt{3}\cos^2\theta\,S_\sigma],$$

and

$$G(e_{3u}),\ = -\frac{1}{2}\sqrt{10}\sin^3\theta\cos\theta\,[\sqrt{3}\,S_\pi + \sqrt{2}\,S_\sigma].$$

It is also possible to carry out the required transformations by the generalised rotation group method (*46*) and to obtain the same results, but with this technique it is rather easier to lose the ubiquitous factor of (– 1).

These equations differ from those of *Fischer* (*7*) in that the relative signs of the S_σ and S_π contributions are here reversed, and consequently the data (*27*) used to calculate the G_{ML} for $U(Cot)_2$ for the a_{2u}, e_{1u}, e_{2u}, and e_{3u} (σ, π, δ, and ϕ) orbitals lead to a $|G_{ML}|$ sequence of

$$\pi < \phi < \sigma < \delta \quad \text{(see Table 1)}$$

as against Fischer's order of

$$\phi < \sigma < \delta < \pi .$$

Clearly this discrepancy is of some consequence, especially since *Amberger et al.* (*15*) used *Fischer's* (*7*) G_{ML} expressions in their recent treatment of the magnetic properties of $U(Cot)_2$. Thus *Amberger et al.* (*15*), commenting on the differences between their numerical G_{ML} values and those of *Hayes* and *Edelstein* (*27*), point out that only the former meet the requirements of Unsöld's theorem (*47*), but as this test involves the sums of the squares of the coefficients of the λ components of G_{ML} it would naturally give no check on the relative signs of S_σ and S_π. Moreover, *Mowery* (*48*) has confirmed the numerical values obtained by *Warren* (*26*), and *Hayes* and *Edelstein* also agree (*27*) in the relative signs of S_σ and S_π. In fact *Hayes* (*49*) has also informed the author that the G_{ML} values of *Hayes* and *Edelstein* (*27*) were based on $\theta = 45°$ (not 43.40°) for $U(Cot)_2$ and that the S_π contributions were too large by a factor of $\sqrt{2}$. When allowance is made for this the results of *Warren* (*26*) and *Hayes* and *Edelstein* (*27*) are in full agreement and both naturally satisfy Unsöld's theorem.

For the calculation of the bonding or anti-bonding energies of the metal f-orbitals it is now however necessary to consider how the one-electron H terms may be estimated, both for the metal and for the ligand symmetry orbitals. In the latter case the H_L terms for the levels of $a(\sigma)$, $e_1(\pi)$, $e_2(\delta)$, and $e_3(\phi)$ symmetry have been estimated using a simple Hückel technique by *Fischer* (*7*), and the values −15, −13, −9, and −5 eV respectively obtained, but the derivation of values for H_M presents rather more difficulty since the necessary ionisation potentials are usually either unknown or highly configuration dependent. However, for the 4 f series the photoelectron spectra of the triscyclopentadiene lanthanide complexes show that the f-orbital ionisation occurs fairly regularly at between about 7 and 7.5 eV for the early members of the series, and the value of −7 eV was therefore adopted by *Warren* (*26*) for the $[Ce(Cot)_2]^-$ system. For the 5f series *Hayes* and *Edelstein* (*27*) adopted, with some misgivings, a value of −8 eV for H_M since a more negative value of −11.3 eV, suggested by a theoretical ionisation potential, would have resulted in the accumulation of far too high an electron population on the metal. However, there seems ample justification for the value of −8 eV since it is approximately equivalent to the experimentally determined first f-orbital ionisation energy for uranium I (*50, 51*), whilst a very recent measurement of the photoelectron spectrum of $U(Cot)_2$ has shown the f-orbital ionisation to occur there at 6.2 eV (*52*), thus suggesting that possibly even −8 eV may be too negative. The latter value was though also adopted by *Warren* (*26*), together with the primitive $2p_\sigma - 5f$ and $2p_\pi - 5f$ overlaps quoted by *Hayes* and *Edelstein* (*27*), whilst for $[Ce(Cot)_2]^-$ the primitive $2p_z - 4f_{z^3}$ and $2p_x - 4f_{xz^2}$ overlaps were calculated (*26*) using the self-consistent field 4f wave functions for Ce^{3+} of *Watson* and *Freeman* (*38*) and the double-zeta 2p carbon orbitals of *Clementi* (*53*). For both the cerium and uranium complexes the group overlaps, G_{ML}, were evaluated as described above, ligand-ligand overlap being neglected throughout.

For both the complexes considered the metal f-orbitals lie, on a one-electron picture, below the ligand orbitals of $e_3(\phi)$ symmetry and above those belonging to the $a(\sigma)$, $e_1(\pi)$, and $e_2(\delta)$ representations. Consequently, as indicated in Section 1, the systems may be pictured as accommodating the 20 electrons of the two cyclo-octatetraenyl dianion rings in the a_{2u}, a_{1g}, e_{1u}, e_{2u}, and e_{2g}, mainly ligand levels, with the one or two f-electrons of Ce^{III} or U^{IV} respectively located in the mainly metal f, e_{3u} level. Moreover, the a_{2u}, e_{1u}, and e_{2u} mainly metal f-levels will all be anti-bonding in character, whilst the e_{3u} f-orbital will be bonding.

Thus the one-electron orbital schemes of both *Warren* (*26*) and of *Hayes* and *Edelstein* (*27*) follow in principle the original suggestions of *Streitwieser* and *his collaborators* (*8, 25*), but *Amberger et al.* (*15*), in fitting their recent magnetic susceptibility results, located the uranium 5f H_M level as lying between the ligand a and e_1 orbitals (i.e. between about – 13 and – 15 eV), this being necessary to produce crystal field splitting parameters which would accommodate their preferred deduction of a $J_z = 0$ ground state component for $U(Cot)_2$ (see Section 3). Bearing

Table 1. One-Electron Molecular Orbital Calculations for $[Ce(Cot)_2]^-$ and $U(Cot)_2$.

f-orbital	G_{ML}	E* (cm^{-1})	Effective Splitting	Eigenvectors c_M	c_L
$[Ce(Cot)_2]^-$: $S_\sigma = 0.0226$, $S_\pi = 0.0149$, $H_M = -7$ eV, $\theta = 40.85°$					
σ	+ 0.0327	+ 243	+ 29.3	0.9981	– 0.0610
π	– 0.0235	+ 125	– 88.1	0.9987	+ 0.0507
δ	– 0.0406	+ 538	+ 324.5	0.9847	+ 0.1742
ϕ	– 0.0194	– 38	– 251.1	0.9988	– 0.0482

Ds = 25.0, Dt = – 39.3, Du = 8.3 cm^{-1} ($B_0^2 = 374.9$, $B_0^4 = -1296.3$, $B_0^6 = 707.9$ cm^{-1}), charge on Ce = + 2.86.

f-orbital	G_{ML}	E* (cm^{-1})	Effective Splitting	Eigenvectors c_M	c_L
$U(Cot)_2$: $S_\sigma = 0.0409$, $S_\pi = 0.0287$, $H_M = -8$ eV, $\theta = 43.40°$					
$\sigma(a_{2u})$	+ 0.0699	+ 1238	+ 27.7	0.9897	– 0.1429
$\pi(e_{1u})$	– 0.0268	+ 195	– 1015.3	0.9976	+ 0.0691
$\delta(e_{2u})$	– 0.0735	+ 3530	+ 2319.7	0.8845	+ 0.4666
$\phi(e_{3u})$	– 0.0401	– 108	– 1318.3	0.9978	– 0.0659

Ds = 85.7, Dt = – 274.4, Du = 66.5 cm^{-1} ($B_0^2 = 1286.1$, $B_0^4 = -9053.6$, $B_0^6 = 5709.1$ cm^{-1}), charge on U = + 3.08.

For both Ce and U complexes $H_L(\sigma) = -15$, $H_L(\pi) = -13$, $H_L(\delta) = -9$, and $H_L(\phi) = -5$ eV.

in mind the caveat of the preliminary SCF MO results for $[Ce(Cot)_2]^-$ (*28*), it is of course possible that the one-electron scheme may be only a convenient artifact with no clear relationship to the actual orbital energies, but the location of H_M in $U(Cot)_2$ below – 13 eV is internally at variance with a Wolfsberg-Helmholtz type treatment since it must of necessity result in much too great a metal 5f population (c.f. *Hayes* and *Edelstein* (*27*)), and is very difficult to reconcile with the observed (*52*) f-orbital ionisation energy in $U(Cot)_2$ of 6.2 eV. The merits of these approaches and the reasons for preferring the Streitwieser scheme will however be discussed further in Section 3.

In Table 1 are given the results of the molecular orbital calculations of *Warren* (*26*) for $[Ce(Cot)_2]^-$ and $U(Cot)_2$, including the group overlap integrals, the bonding or anti-bonding energies, the effective splittings, and the eigenvectors for the various levels. (Note that the listed B_0^4 parameter for $U(Cot)_2$ corrects a typographical error in reference (*26*).) These results will be treated more fully in Section 3, and used to make predictions for other 4f and 5f biscyclooctatetraene complexes. The effective splittings listed are simply obtained from the ΔE values by requiring the former to preserve a zero centre of gravity for the f-orbital set, and the principal features which emerge from this molecular orbital model are that the $e_{3u}(\phi)$ level is predicted to be slightly bonding and the $e_{2u}(\delta)$ level to be rather more strongly anti-bonding, whilst the $a_{2u}(\sigma)$ and $e_{1u}(\pi)$ orbitals lie between the ϕ and δ levels and are but weakly anti-bonding.

Attention may now therefore be turned away from the manner in which numerical values of the splitting parameters may be calculated and directed instead towards the application of the ligand field model to systems of pseudo-axial symmetry, and the results of such a treatment, the f^1 configuration being conveniently treated as a first example. In this case the single f-electron produces only a ^2F free ion spectroscopic state which under the influence of spin-orbit coupling yields the two J levels, $^2F_{5/2}$ and $^2F_{7/2}$, the former lying lower by $7/2\ \xi$. On application of a pseudo-axial, C_∞^*, field these levels are split to yield

$$^2F_{5/2} \rightarrow \Pi^* + \Delta^* + \Phi^*$$

and

$$^2F_{7/2} \rightarrow \Pi^* + \Delta^* + \Phi^* + \Gamma^*$$

so that to evaluate the appropriate matrix elements of V_{ax} it is necessary to obtain the relevant Π^*, Δ^*, Φ^*, and Γ^* wave functions. However, as noted earlier, these representations correspond respectively to the M_J (or J_z) values ± 1/2, ± 3/2, ± 5/2, and ± 7/2, so that all that is needed is to write down the trivial $|LSM_LM_S\rangle$ set and to derive the requisite $|LSJM_J\rangle$ set therefrom. The matrix elements of V_{ax} are then easily found, and since the free ion spin-orbit terms are well known the complete perturbation matrices may be written down as in Table 2. For the diagonal elements the results are of course equivalent to those obtained in terms of the $A_k^q\langle r^k\rangle$ parameters by the operator equivalent method of *Stevens* (*32, 33*).

Table 2. Weak Field Matrix Elements of V_{ax} for f^1, C_∞^*, Systems.

Representation $\Pi^*(M_J = \pm 1/2)$			
	$^2F_{5/2}$	$^2F_{7/2}$	
$^2F_{5/2}$	2 (12 Ds + 11 Dt)	$-2\sqrt{3}$ (Ds + 5 Dt + 35 Du)	$\cdot\frac{1}{7}$
$^2F_{7/2}$		(25 Ds + 27 Dt + 35 Du)	
Representation $\Delta^*(M_J = \pm 3/2)$			
	$^2F_{5/2}$	$^2F_{7/2}$	
$^2F_{5/2}$	(6 Ds − 33 Dt)	$-\sqrt{10}$ (3 Ds + 8 Dt − 9 Du)	$\cdot\frac{1}{7}$
$^2F_{7/2}$		3 (5 Ds − 3 Dt − 21 Du)	
Representation $\Phi^*(M_J = \pm 5/2)$			
	$^2F_{5/2}$	$^2F_{7/2}$	
$^2F_{5/2}$	(− 30 Ds + 11 Dt)	$-\sqrt{6}$ (5 Ds − 10 Dt + 7 Du)	$\cdot\frac{1}{7}$
$^2F_{7/2}$		(− 5 Ds − 39 Dt + 35 Du)	
Representation $\Gamma^*(M_J = \pm 7/2)$		Diagonal Spin-Orbit Energies	
	$^2F_{7/2}$	$^2F_{5/2} = -2\xi$	
$^2F_{7/2}$	(− 5 Ds + 3 Dt − Du)	$^2F_{7/2} = +3\xi/2$	

For the f^2 configuration the problem is a little more difficult since now the matrix elements of the electron repulsion operator, e^2/r_{12}, are also involved. However, the necessary $|LSM_LM_S\rangle$ wave functions for the f^2 spectroscopic states have been given by *Satten* and *Margolis* (*54*), together with the matrix elements of spin-orbit coupling. From the full set of $|LSM_LM_S\rangle$ functions the required $|LSJM_J\rangle$ functions may be obtained as before and the matrix elements of V_{ax} evaluated, and the results for the f^2 configuration have also been given by *Warren* (*26*), in terms of the σ, π, δ, and ϕ parameters defined earlier. For f^3 and higher configurations the evaluation of the matrix elements of V_{ax} via the $|LSJM_J\rangle$ wave functions becomes intolerably tedious, and for these systems the results are best obtained using the tensor operator technique described by *Wybourne* (*35*). However, for the interpretation of magnetic properties for example it is frequently possible to derive useful results using simply the splitting of the known weak field ground state indicated by the diagonal matrix elements of V_{ax}, and these latter can still be derived very easily using *Stevens'* (*32*) operator equivalent method, the necessary numerical quantities being also tabulated by (amongst others) *Hutchings* (*36*) and *Abragam* and *Bleaney* (*34*). In this manner the diagonal matrix elements of V_{ax} were found by *Warren* (*26*) for f^x ground states (x = 1–4) in terms of the Ds, Dt, and Du quantities, and used to describe the magnetic susceptibilities of the corresponding sandwich species (see Section 3). (Note however that in the listed values for the f^2, $^3H_4$, ground state in reference (*26*) the vertical order of the numerical entries in the Ds, Dt, and Du columns should be reversed).

3. Interpretation of Magnetic Susceptibilities and Other Experimental Data

For the f-orbital sandwich complexes the application of the ligand field model to the interpretation of experimental data is hindered by the almost complete absence of results from which reliable empirical splitting parameters (B_0^n values) can be derived. Thus electronic spectral data are as yet very limited and although bands interpreted as f–f transitions have been observed, both in the 4f and the 5f series (vide infra), it has not so far proved possible to analyse these in terms of a pseudo-axial, C_∞^*, field in order to afford the desired B_0^n quantities. Moreover, the crucially important f^1 systems – $[Ce(Cot)_2]^-$ and $Pa(Cot)_2$ – whose electronic spectra might most readily be unequivocably interpreted, have not been studied at all from this aspect. For the cerium complex however the single free ion f–f transition, $^2F_{5/2} \rightarrow {}^2F_{7/2}$, is expected to occur between 2 and 2.5 kK (*55*), below the readily accessible energy range, and would in any case probably be obscured by vibrational transitions, whilst for $Pa(Cot)_2$, in which the band should be found in the more promising 5 to 7 kK region, the experimental difficulties of preparation and manipulation are probably greater than for any of the other compounds.

Moreover, no esr experiments have been reported for any of the $M(Cot)_2$ species, and this is especially unfortunate since studies of the Kramer's doublet f^1 and f^3 systems would be of the greatest value in giving independent evidence about the nature of the lowest lying C_∞^* components of the ground levels. Similarly, although the photoelectron spectrum of $U(Cot)_2$ has been measured (*52*), it was not possible to resolve the f-ionisation peak and consequently again no further information about the ground state could be deduced. Thus there remains only the magnetic susceptibility data, which are often hard to interpret definitively, but attention is now directed towards testing the correlation between this evidence and the ligand field model: this will therefore occupy the major part of this Section, after which a brief survey will be given of relevant results from other techniques (e.g. uv spectra, nmr, pes, etc.).

It is of course well known that magnetic moment values and their temperature dependence do not usually allow definitive values of the crystal field splitting parameters to be determined. This is because the magnetic susceptibility is a bulk property dependent upon a thermal Boltzman population of the various sub-levels which constitute the ground state (and sometimes upon higher states as well), and generally there exist more unknown parameters than can be uniquely defined by the experimental data. Consequently the procedure now adopted will first of all calculate the ground state splittings of the various f^x configurations of the 4f and 5f series, using the B_0^n values obtained for the $[Ce(Cot)_2]^-$ and $U(Cot)_2$ complexes (Table 1) as being representative for the respective lanthanide and actinide series. The expressions for the magnetic moments will then be derived ($\mu_\parallel$ and $\mu_\perp$), which involve the calculated energy splittings of the ground states, and the temperature dependence of the moment expressed in terms of the parameter kT/Δ, where Δ is the overall ground state splitting. In this way the μ vs kT/Δ plots will have the correct form even if the actual value of Δ is incorrect, as long as the ratio of the B_0^2, B_0^4, and B_0^6 values remains unchanged. Some estimate of the validity of the ligand field approach may then be obtained by assessing how well the predicted magnetic behaviour accords with the experimental results.

Initially the f^1, f^2, f^3, and f^4 configurations will be treated, these having the free ion ground states $^2F_{5/2}$, $^3H_4$, $^4I_{9/2}$, and $^5I_4$ respectively, and the first order ground state splittings in terms of the crystal field parameters have been given (*26*), using the operator equivalent method (*32*). The use of the diagonal terms only may be justified by the fact that J mixing due to the ligand field is found (vide infra) to be small and to have little effect on the moment. It is now necessary to calculate the matrix elements of the Zeeman operators within the respective ground states, and again the first order $|LSJM_J\rangle$ sets are used, thus neglecting the effects of the mixing of states due to intermediate coupling. For the 4f series this is a rather good approximation since J is generally a good quantum number and the ground state purity is never less than about 97%, but because of the stronger intermediate coupling somewhat lower purities are found for the 5f series, ranging from about 95% for $5f^2$ to only 81% for $5f^4$.

In the 4f series the respective free ion ground states are separated by not less than 2000 cm^{-1} from the first excited state, and since this gap is large compared with kT the first order ground state manifold wave functions should suffice to give an excellent account of the susceptibilities. In the 5f series the gap between the lowest J level and the first excited state should be greater, due to the larger spin-orbit coupling, but because of the stronger spin-orbit mixing the first order ground state functions will not give such a good account of the magnetic properties, especially for the $5f^3$ and $5f^4$ systems (vide supra).

For the calculation of the Zeeman matrix elements it is convenient to introduce the orbital reduction factor, k′, in an attempt to allow for possible delocalisation of electrons out of the f-orbitals onto the ligand atoms. Thus the elements of $k'L_z + 2S_z$ and of $k'L_x + 2S_x$ were determined within the various ground state manifolds, assuming for simplicity a single isotropic value of k′. This approximation is however less satisfactory than for the sandwich complexes of the 3d series since there the wave functions correspond closely to pure strong field configurations and for orbitally degenerate ground states the terms involving k′ are restricted to the L_z operator and relate only to the degenerate level actually making the orbital contribution. In the present weak field basis however k′ terms occur for both L_z and $L_x(L_y)$ and the $|LSJM_J\rangle$ states are in general functions of all the $|m_l|$ values. Thus a single isotropic k′ neglects differences in metal-ligand mixing for the real f-orbitals of differing $|m_l|$, and the derived values of k′ must therefore be somewhat sceptically interpreted. The approximation is however useful since it avoids the necessity of adopting four different reduction parameters relating to the various $|m_l|$ values.

For the f^1 and f^2 systems the required Zeeman elements are easily found from the known $|LSJM_J\rangle$ functions, and for all configurations a general formula for the case $k' = 1$ has been given by *Wybourne* (*35*). The results for all f^x configurations are however easily found for $k' \neq 1$, via the Clebsch-Gordan coefficients, using the expressions given in Table 3, wherein are also listed the resulting Zeeman matrix ele-

Table 3. Matrix Elements of Zeeman Operators within f^x Ground States.

Elements are listed in terms of the appropriate $J_z(M_J)$ values: all unlisted elements are zero.

$f^1, {}^2F_{5/2}$

$k'L_z + 2S_z$: $\langle \pm J_z \| \pm J_z \rangle = \pm J_z (2/7) \cdot (4k' - 1)$

$k'L_x + 2S_x$: Multiply all entries by $(1/7) \cdot (4k' - 1)$

$\langle \pm 5/2 \| \pm 3/2 \rangle = \sqrt{5}$; $\langle \pm 3/2 \| \pm 1/2 \rangle = 2\sqrt{2}$; $\langle \pm 1/2 \| \mp 1/2 \rangle = 3$.

$f^2, {}^3H_4$

$k'L_z + 2S_z$: $\langle \pm J_z \| \pm J_z \rangle = \pm J_z (2/5) \cdot (3k' - 1)$

$k'L_x + 2S_x$: Multiply all entries by $(1/5) \cdot (3k' - 1)$

$\langle \pm 4 \| \pm 3 \rangle = 2\sqrt{2}$; $\langle \pm 3 \| \pm 2 \rangle = \sqrt{14}$; $\langle \pm 2 \| \pm 1 \rangle = 3\sqrt{2}$; $\langle \pm 1 \| 0 \rangle = 2\sqrt{5}$.

$f^3, {}^4I_{9/2}$

$k'L_z + 2S_z$: $\langle \pm J_z \| \pm J_z \rangle = \pm J_z (2/11) \cdot (7k' - 3)$

$k'L_x + 2S_x$: Multiply all entries by $(1/11) \cdot (7k' - 3)$

$\langle \pm 9/2 \| \pm 7/2 \rangle = 3$; $\langle \pm 7/2 \| \pm 5/2 \rangle = 4$; $\langle \pm 5/2 \| \pm 3/2 \rangle = \sqrt{21}$; $\langle \pm 3/2 \| \pm 1/2 \rangle = 2\sqrt{6}$;
$\langle \pm 1/2 \| \mp 1/2 \rangle = 5$.

$f^4, {}^5I_4$

$k'L_z + 2S_z$: $\langle \pm J_z \| \pm J_z \rangle = \pm J_z (1/5) \cdot (7k' - 4)$

$k'L_x + 2S_x$: Multiply all entries by $(1/10) \cdot (7k' - 4)$

$\langle \pm 4 \| \pm 3 \rangle = 2\sqrt{2}$; $\langle \pm 3 \| \pm 2 \rangle = \sqrt{14}$; $\langle \pm 2 \| \pm 1 \rangle = 3\sqrt{2}$; $\langle \pm 1 \| 0 \rangle = 2\sqrt{5}$.

When the functions are defined in terms of the quantum numbers J and J_z the entries of Table 3 and similar related results may be obtained using the expressions: –

$$\langle J, J_z | k'L_z + 2S_z | J, J_z \rangle = \sum_{M_L, M_S} \langle LS M_L M_S | LSJ J_z \rangle^2 \cdot \{k'M_L + 2M_S\}$$

$$\langle J, J_z | k'L_z + 2S_z | J', J_z \rangle = \sum_{M_L, M_S} \delta_{M_L, M'_L} \delta_{M_S, M'_S} \cdot$$
$$\langle LS M_L M_S | LSJ J_z \rangle \cdot \langle LS M'_L M'_S | LSJ' J_z \rangle \cdot \{k'M_L + 2M_S\} \quad (J' = J \pm 1)$$

$$\langle J, J_z | k'L_x + 2S_x | J', J_z \pm 1 \rangle = \sum_{M_L, M_S} [\langle LS M_L M_S | LSJ J_z \rangle \cdot$$
$$\langle LS M_L \pm 1\, M_S | LSJ' J_z \pm 1 \rangle \cdot \frac{1}{2} k' \{(L \pm M_L + 1)(L \mp M_L)\}^{1/2} + \langle LS M_L M_S | LSJ J_z \rangle \cdot$$
$$\langle LS M_L M_S \pm 1 | LSJ' J_z \pm 1 \rangle \{(S \pm M_S + 1)(S \mp M_S)\}^{1/2}] \quad (J' = J \text{ or } J \pm 1).$$

ments within the various ground states. The elements of $H_\parallel(k'L_z + 2S_z)$ are entirely diagonal and those of $H_\perp(k'L_x + 2S_x)$ entirely off-diagonal, and for the latter the eigenvalues are obtained by second order perturbation theory as previously described (*56*). The susceptibility is then found via the well known Van Vleck equation

$$\chi_\alpha = \frac{N \sum_i [(E_i^{I\,2}/kT) - 2\,E_i^{II}]\, e^{-E_i^0/kT}}{\sum_i e^{-E_i^0/kT}}$$

where

$$E_i^I = \langle \psi_i | k'L_\alpha + 2\,S_\alpha | \psi_i \rangle \beta ,$$

and

$$E_i^{II} = \sum_j (\langle \psi_i | k'L_\alpha + 2\,S_\alpha | \psi_j \rangle \beta)^2/(E_i^0 - E_j^0)$$

with $i \neq j$ and $\alpha = z$ ($H_\parallel$) or x, y ($H_\perp$). The effective magnetic moments are then derived from the Curie formula $\chi = N\,\mu_{eff}^2/3\,kT$, and the average, from the z and x, y components using $\langle\mu\rangle = [1/3\,(\mu_x^2 + \mu_y^2 + \mu_z^2)]^{1/2}$. The expressions for $\mu_\parallel$ and $\mu_\perp$ thus derived are listed in Table 4, and the calculated temperature dependence of the moments shown in Figs. 6 to 9.

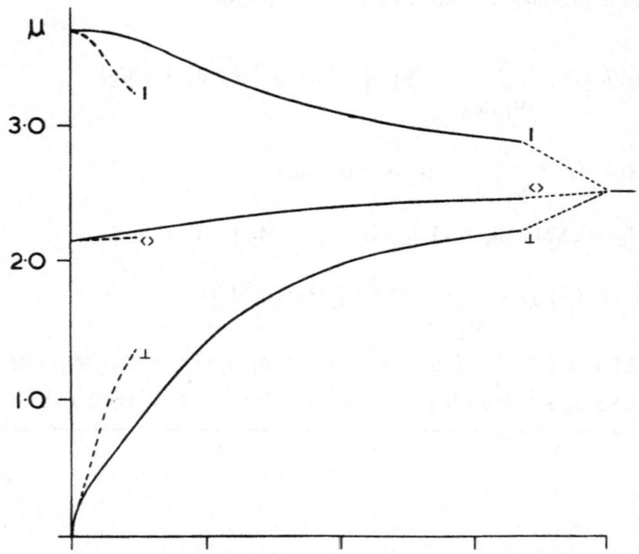

Fig. 6. Temperature Dependence of Magnetic Moments for f^1 C_∞^* Systems. (Here and in Figs. 7–9, —— 4f, - - - - - 5f)

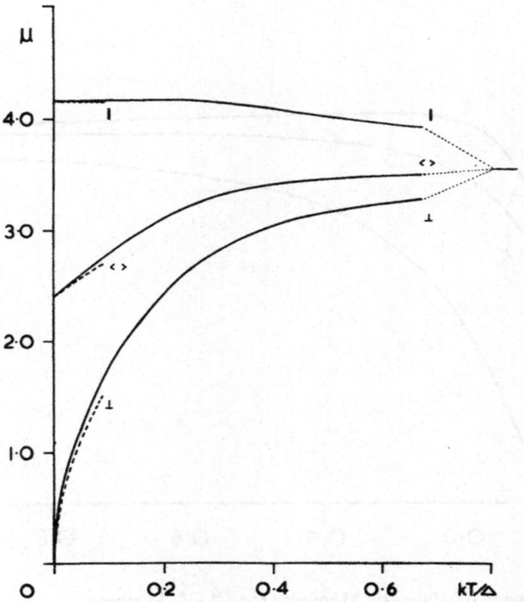

Fig. 7. Temperature Dependence of Magnetic Moments for f^2 C^*_∞ Systems.

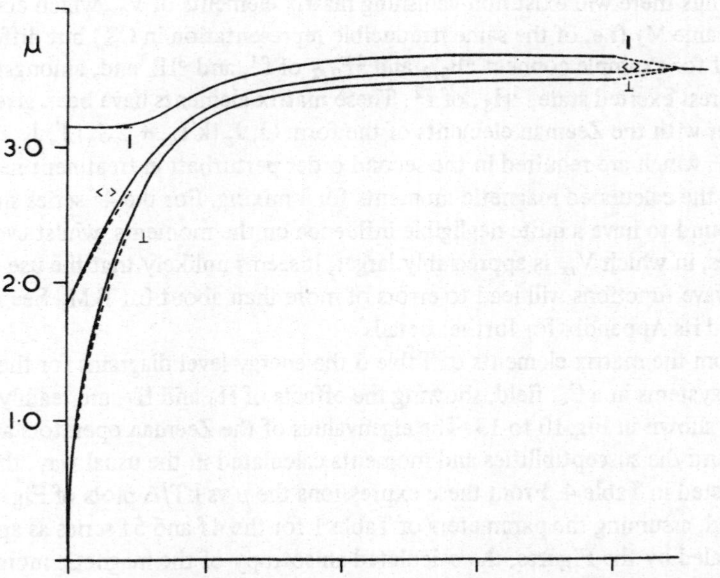

Fig. 8. Temperature Dependence of Magnetic Moments for f^3 C^*_∞ Systems.

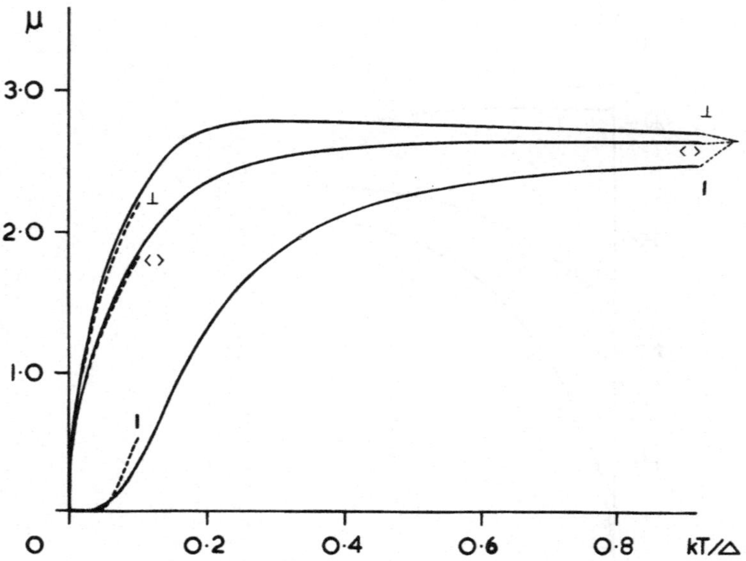

Fig. 9. Temperature Dependence of Magnetic Moments for f^4 C^*_∞ Systems.

Quite apart from the question of the quality of the first order $|LSJM_J\rangle$ functions of the ground state in the free ion situation there arises the problem of the extent to which the values of the moment may be influenced by J mixing due to the ligand field. Thus there will exist non-vanishing matrix elements of V_{ax} which connect states of the same M_J (i.e. of the same irreducible representation in C^*_∞) but different J, and will for example connect $^2F_{5/2}$ and $^2F_{7/2}$ of f^1, and $^3H_4$ and, amongst others, the nearest excited state, $^3H_5$, of f^2. These matrix elements have been given (*26*), together with the Zeeman elements of the form $\langle J, J_z | k'L_\alpha + 2 S_\alpha | J', J_z \pm 1\rangle$, $J' = J$ or $J \pm 1$, which are required in the second order perturbation treatment needed to correct the calculated magnetic moments for J mixing. For the 4f series such effects were found to have a quite negligible influence on the moments, whilst even for the 5f series, in which V_{ax} is appreciably larger, it seems unlikely that the use of the first order wave functions will lead to errors of more than about 0.1 B.M.. See reference (*26*) and its Appendix for further details.

From the matrix elements of Table 3 the energy level diagrams for the f^1, f^2, f^3, and f^4 systems in a C^*_∞ field, showing the effects of $H_\parallel$ and $H_\perp$, are readily derived and are shown in Fig. 10 to 13. The eigenvalues of the Zeeman operators are then found and the susceptibilities and moments calculated in the usual way, the results being listed in Table 4. From these expressions the μ vs kT/Δ plots of Fig. 6 to 9 are obtained, assuming the parameters of Table 1 for the 4f and 5f series as appropriate. As revealed by the Figures, the calculated anisotropy of the magnetic moment is appreciable for all the f^x configurations treated, and the crystallographic data (*22, 23*) for $[Ce(Cot)_2]^-$ and $U(Cot)_2$ show that both belong to monoclinic space groups: it should therefore be possible to measure this anisotropy experimentally although the

Table 4. Magnetic Moments for f^x Systems in C^* Symmetry.

f^1, $^2F_{5/2}$: $\alpha = (\Delta^* - \Phi^*)/kT$, $\beta = (\Pi^* - \Phi^*)/kT$

$$\mu_\parallel = \frac{\sqrt{3}\,(4\,k' - 1)}{7}\left\{\frac{25 + 9\,e^{-\alpha} + e^{-\beta}}{1 + e^{-\alpha} + e^{-\beta}}\right\}^{1/2}$$

$$\mu_\perp = \frac{\sqrt{3}\,(4\,k' - 1)}{7}\{[10\,(1 - e^{-\alpha})\,\alpha^{-1} + 16\,(e^{-\beta} - e^{-\alpha})\,(\alpha - \beta)^{-1} + 9\,e^{-\beta}]\cdot [1 + e^{-\alpha} + e^{-\beta}]^{-1}\}^{1/2}$$

f^2, $^3H_4$: $\alpha = (\Gamma - \Phi)/kT$, $\beta = (\Delta - \Phi)/kT$, $\gamma = (\Pi - \Phi)/kT$, $\delta = (\Sigma^+ - \Phi)/kT$

$$\mu_\parallel = \frac{2\sqrt{6}\,(3\,k' - 1)}{5}\left\{\frac{9 + 16\,e^{-\alpha} + 4\,e^{-\beta} + e^{-\gamma}}{e^{-\delta} + 2\,(1 + e^{-\alpha} + e^{-\beta} + e^{-\gamma})}\right\}^{1/2}$$

$$\mu_\perp = \frac{2\sqrt{6}\,(3\,k' - 1)}{5}\{[4\,(1 - e^{-\alpha})\,\alpha^{-1} + 7\,(1 - e^{-\beta})\,\beta^{-1} + 9\,(e^{-\beta} - e^{-\gamma})\,(\gamma - \beta)^{-1} + 10\,(e^{-\gamma} - e^{-\delta})\,(\delta - \gamma)^{-1}]\cdot [e^{-\delta} + 2\,(1 + e^{-\alpha} + e^{-\beta} + e^{-\gamma})]^{-1}\}^{1/2}$$

f^3, $^4I_{9/2}$: $\alpha = (H^* - \Phi^*)/kT$, $\beta = (\Gamma^* - \Phi^*)/kT$, $\gamma = (\Delta^* - \Phi^*)/kT$, $\delta = (\Pi^* - \Phi^*)/kT$.

$$\mu_\parallel = \frac{\sqrt{3}\,(7\,k' - 3)}{11}\left\{\frac{25 + 81\,e^{-\alpha} + 49\,e^{-\beta} + 9\,e^{-\gamma} + e^{-\delta}}{1 + e^{-\alpha} + e^{-\beta} + e^{-\gamma} + e^{-\delta}}\right\}^{1/2}$$

$$\mu_\perp = \frac{\sqrt{3}\,(7\,k' - 3)}{11}\{[32\,(1 - e^{-\beta})\,\beta^{-1} + 42\,(1 - e^{-\gamma})\,\gamma^{-1} + 18\,(e^{-\beta} - e^{-\alpha})\,(\alpha - \beta)^{-1} + 48\,(e^{-\gamma} - e^{-\delta})\,(\delta - \gamma)^{-1} + 25\,e^{-\delta}]\cdot [1 + e^{-\alpha} + e^{-\beta} + e^{-\gamma} + e^{-\delta}]^{-1}\}^{1/2}$$

f^4, $^5I_4$: $\alpha = (\Gamma - \Sigma^+)/kT$, $\beta = (\Phi - \Sigma^+)/kT$, $\gamma = (\Delta - \Sigma^+)/kT$, $\delta = (\Pi - \Sigma^+)/kT$

$$\mu_\parallel = \frac{\sqrt{6}\,(7\,k' - 4)}{5}\left\{\frac{16\,e^{-\alpha} + 9\,e^{-\beta} + 4\,e^{-\gamma} + e^{-\delta}}{1 + 2\,(e^{-\alpha} + e^{-\beta} + e^{-\gamma} + e^{-\delta})}\right\}^{1/2}$$

$$\mu_\perp = \frac{\sqrt{6}\,(7\,k' - 4)}{5}\{[4\,(e^{-\alpha} - e^{-\beta})\,(\beta - \alpha)^{-1} + 7\,(e^{-\beta} - e^{-\gamma})\,(\gamma - \beta)^{-1} + 9\,(e^{-\delta} - e^{-\gamma})\,(\gamma - \delta)^{-1} + 10\,(1 - e^{-\delta})\,\delta^{-1}]\cdot [1 + 2\,(e^{-\alpha} + e^{-\beta} + e^{-\gamma} + e^{-\delta})]^{-1}\}^{1/2}.$$

extreme instability of all the An and Ln Cot complexes renders this a formidable undertaking, but as yet, although such data would be of immense value, only average moment values are available.

It is nevertheless of interest to consider the predicted temperature dependence of both $\mu_\parallel$ and $\mu_\perp$ for the various f^x systems since such data would afford much more definitive evidence concerning the lowest lying ligand field components of the various ground states than do the average moment values alone. Thus in Table 5 are listed the lowest ground state components for the f^x ($x = 1$–4) configurations, together with the isotropic free ion value, given by $\mu = g\,(J(J + 1))^{1/2}$, the g value being that appropriate to the free ion J ground state, and the values of $\mu_\parallel$, $\mu_\perp$, and $\langle\mu\rangle$ as $kT/\Delta \to 0$, assuming the B_0^n values of Table 1 and the ensuing $M_J(J_z)$ ordering. The values as $kT/\Delta \to 0$ depend of course on the B_0^n quantities and apply only for the particular M_J component predicted to lie lowest, but it is nonetheless useful that the value of the average moment, $\langle\mu\rangle$, as $kT/\Delta \to 0$ is given by $\langle\mu\rangle = g\,|M_J|$, thereby giving an immediate pointer to the proper $|M_J|$ value.

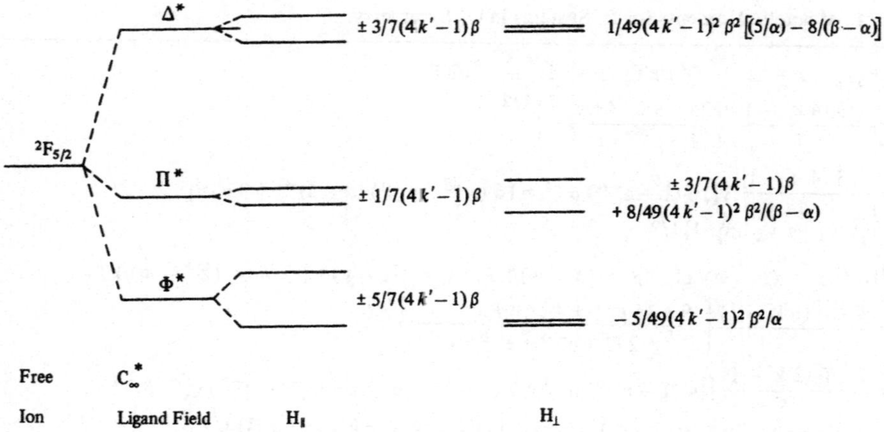

Fig. 10. Energy Level Diagram for f[1] C_∞^* Systems. (Here and in Figs. 11–13 see Table 4 for definitions 1).

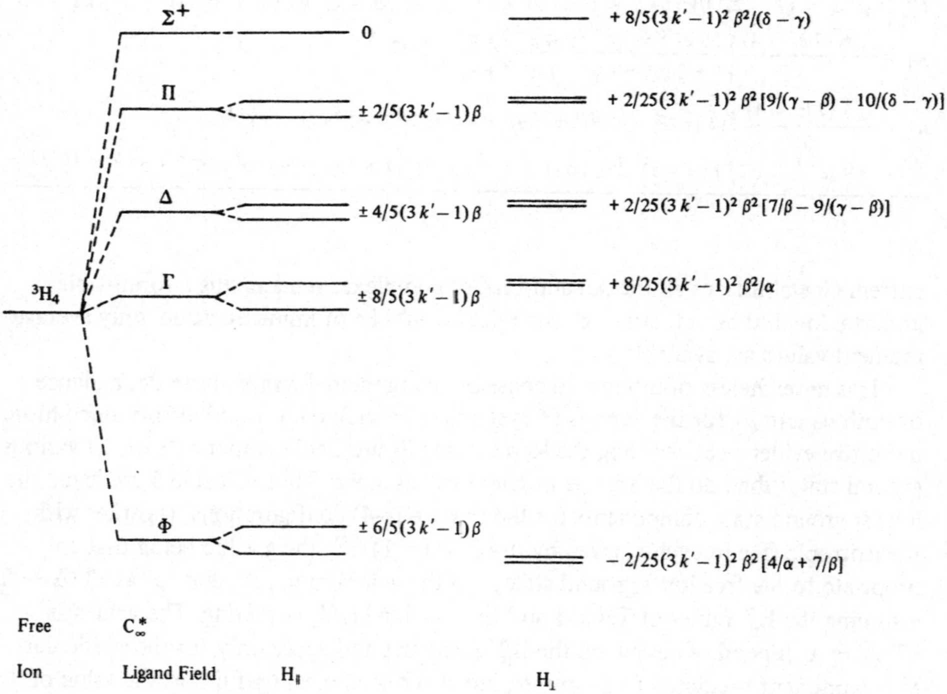

Fig. 11. Energy Level Diagram for f[2] C_∞^* Systems.

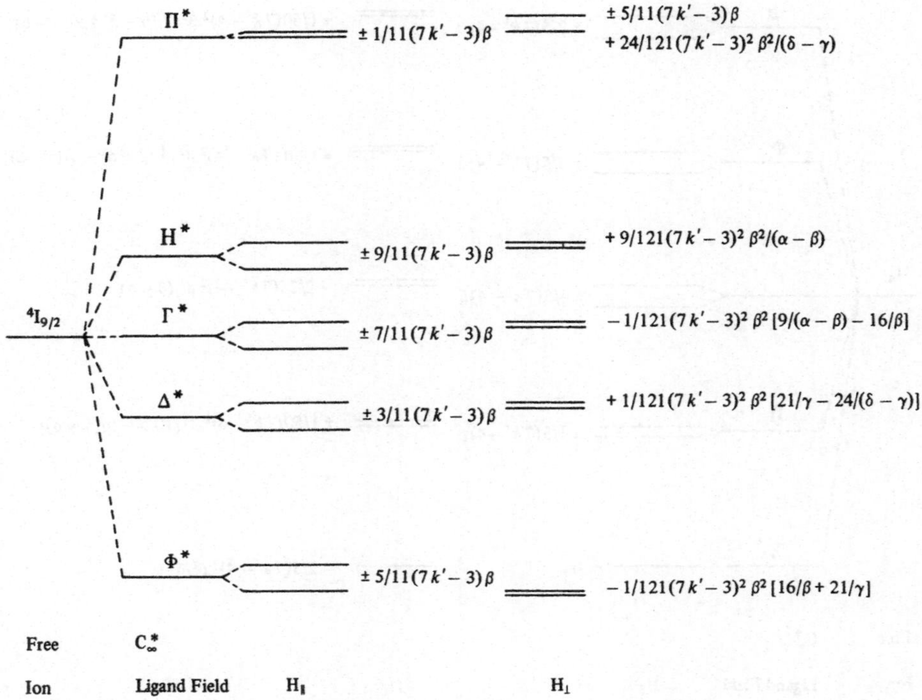

Fig. 12. Energy Level Diagram for f^3 C^*_∞ Systems.

With the B^n_0 values of Table 1 the $\langle\mu\rangle$ value for the f^1 system shows little dependence on temperature, and although $\langle\mu\rangle$ varies somewhat more for the f^2 configuration its small overall temperature variation is in both cases largely due to the contrary temperature dependence of $\mu_\parallel$ and $\mu_\perp$, thus underlining the potential utility of anisotropic measurements. For f^3 and more particularly f^4 systems both $\mu_\parallel$ and $\mu_\perp$ decrease with decreasing temperature, thereby producing a much larger total variation of $\langle\mu\rangle$, but because of the much larger splittings (B^n_0 values) expected for the 5f series the μ vs kT/Δ plots show considerable differences between the actinide and lanthanide series.

Table 5. Limiting Calculated Magnetic Moments for f^x M(Cot)$_2$ Systems.

Configuration and Ground State		Free Ion Value (k′ = 1)	kT/Δ → 0 $\mu_\parallel$	$\mu_\perp$	$\langle\mu\rangle$
f^1, $^2F_{5/2}$	Φ* (± 5/2)	$3\sqrt{5}/\sqrt{7}$	$15\sqrt{3/7}$	0	15/7
f^2, $^3H_4$	Φ (± 3)	$8/\sqrt{5}$	$12\sqrt{3/5}$	0	12/5
f^3, $^4I_{9/2}$	Φ* (± 5/2)	$12/\sqrt{11}$	$20\sqrt{3/11}$	0	20/11
f^4, $^5I_4$	Σ^+ (0)	$6/\sqrt{5}$	0	0	0

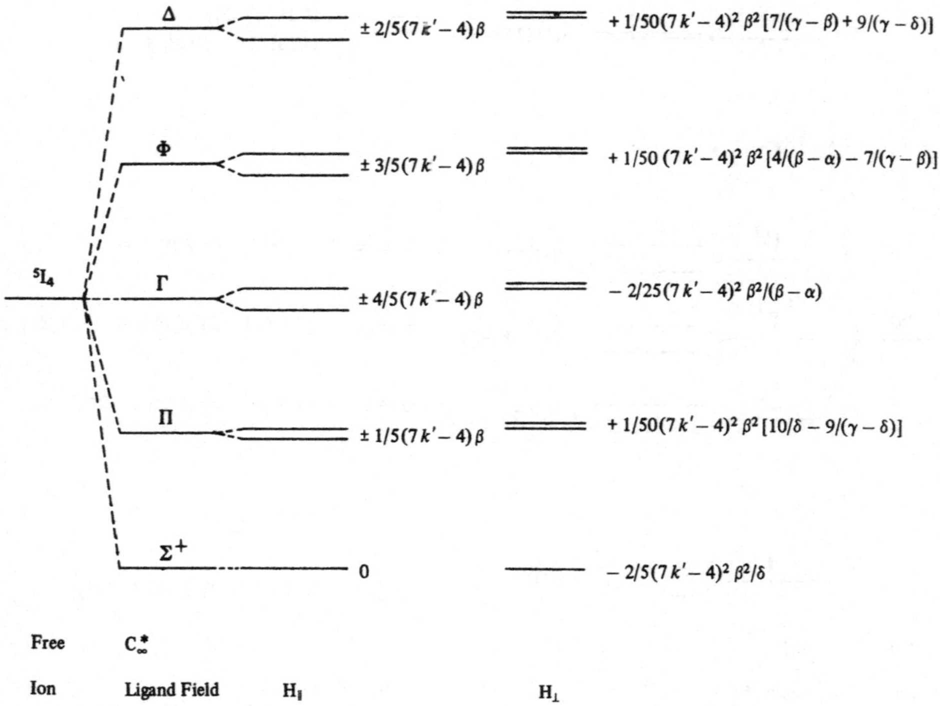

Fig. 13. Energy Level Diagram for f^4 C_∞^* Systems.

For the $[Ln(Cot)_2]^-$ complexes of the 4f series there exist only room temperature measurements of the magnetic moments, apparently relating to THF solutions (*20, 21*), and values have been reported for the Ce, Pr, Nd, Sm, Gd, and Tb compounds. Of these the Gd^{III} salt should, to the first order, show a $^8S_{7/2}$ ground state with an isotropic J-only moment, and is not therefore discussed further. Moreover, for the f^5 Sm^{III} complex a simple treatment along the lines of the present approach would be inadequate since in the free ion the first excited level, the $^6H_{7/2}$ state, lies only about 1000 cm^{-1} above the $^6H_{5/2}$ ground level: thus the value of 1.42 B.M. observed for $[Sm(Cot)_2]^-$, although lying close to the range commonly observed for Sm^{III} compounds (ca. 1.5–1.6 B.M.), may conceal a very complex magnetic situation about which the available evidence does not justify further speculation. Finally, although the $4f^8$, $^7F_6$, Tb^{III} complex is amenable to the present approach, the C_∞^* field splits its ground level into no fewer than seven components, so that results based on the somewhat arbitrary parameters of Table 1 are rather insecure. Treatment of this system is therefore deferred, but in addition to consideration of the data for the Ce, Pr, and Nd complexes (f^1, f^2, and f^3) predictions are also made for the $4f^4$ Pm^{III} system since although it is yet unprepared the synthetic method appears to be of general application and the calculated magnetic properties are especially interesting.

For the Ce, Pr, and Nd complexes respectively *Streitwieser et al.* (*20, 21*) report room temperature average moments of 1.88, 2.84, and 2.98 B.M.. However, although these authors maintained that the observed moments corresponded closely to the free ion quantities, the $\langle\mu\rangle$ values for the Ce, Pr, and Nd species are respectively 0.65, 0.74, and 0.64 B.M. lower than the J-only results and some 0.5–0.6 B.M. lower than the values quoted (*21*) for $Ln(SO_4)_3$, 8 H_2O salts. Consequently one may reasonably conclude that these discrepancies are significant, and the three systems are now therefore considered individually.

For $[Ce(Cot)_2]^-$ the observed $\langle\mu\rangle$ of 1.88 B.M. is the most difficult to interpret. Thus for a $\Phi^*(\pm 5/2)$ lowest component of the $^2F_{5/2}$ ground state the limiting value of $\langle\mu\rangle$ as $kT/\Delta \rightarrow 0$ is actually 2.143 B.M., and to accommodate the measured $\langle\mu\rangle$ value would require an orbital reduction factor, k′, at least as small as 0.91, and probably a substantially larger ligand field as well. The degree of metal-ligand mixing thus implied is substantially greater than that indicated by the results of Table 1, and is also at variance with the markedly ionic chemical properties of the $[Ln(Cot)_2]^-$ species (*21*), and with the orbital reduction factors indicated by esr studies (*57, 58*) of other lanthanide systems. The possibility of other M_J components lying lower than the $\Phi^*(\pm 5/2)$ level cannot be entirely discounted, but were either a $\Pi^*(\pm 1/2)$ or a $\Delta^*(\pm 3/2)$ level to lie lowest the limiting value of $\langle\mu\rangle$ as $kT/\Delta \rightarrow 0$ would be smaller than the observed moment and the divergence between theory and experiment even less readily understood.

For the $4f^2$ and $4f^3$ Pr and Nd complexes the observed room temperature moments are greater than the limiting ($kT/\Delta \rightarrow 0$) values calculated (2.40 and 1.818 B.M. respectively), but still appreciably less than the values predicted by the parameters of Table 1. Thus these latter predict room temperature values of about 3.4–3.5 B.M. for the Pr complex and 3.5–3.6 B.M. for the Nd compound as against the experimental values of 2.84 and 2.98 B.M. respectively, so that either an appreciably larger, ligand field or an orbital reduction factor of as little as about 0.9, or some combination of both, would be required to obtain satisfactory accordance.

Since none of these possibilities appears particularly attractive there remains to consider the likelihood of distortion from the eight-fold, pseudo-axial, symmetry. From the point of view of distortions of Jahn-Teller origin only the f^2 system, with a doubly degenerate $\Phi(\pm 3)$ lowest level, has a ground state which could be split by descent in symmetry; the f^1 and f^3 systems both have Kramers' doublet lowest components, whilst the f^4 configuration is predicted to yield a non-degenerate $\Sigma^+(M_J = 0)$ ground level. In the 4f series however there is no experimental evidence to support the idea of a low symmetry ligand field contribution and it is also a possibility that solvation by THF molecules, which is well established for the $[Ln(Cot)_2]^-$ species (*21*), may play a part by disturbing the effective pseudo-axial symmetry. For the lanthanide series therefore more experimental data are essential, preferably on solid, unsolvated, samples, and extending to the lowest temperatures, whilst esr measurements on the f^1 and f^3 systems would be very valuable, especially in the latter case for which the Nd hyperfine coupling constants might also be obtained.

Table 6. Calculated Ligand Field Splittings for f^x Systems in C_∞^* Symmetry, using the Parameters of Table 1.

	E(4 f)[a]	E(5 f)[b]
f^1, $^2F_{5/2}$		
Component		
Φ^*	− 168.8	− 798.6
Δ^*	+ 206.6	+ 1366.9
Π^*	− 37.8	− 568.3
	Δ = 375.5	Δ = 2165.5
f^2, $^3H_4$		
Component		
Γ	+ 3.3	+ 429.8
Φ	− 235.4	− 1608.3
Δ	+ 27.8	+ 162.7
Π	+ 134.6	+ 706.1
Σ^+	+ 139.3	+ 619.6
	Δ = 374.7	Δ = 2314.4
f^3, $^4I_{9/2}$		
Component		
H^*	+ 2.5	+ 144.4
Γ^*	− 8.4	+ 93.1
Φ^*	− 144.8	− 1128.8
Δ^*	− 17.3	− 252.5
Π^*	+ 168.0	+ 1143.8
	Δ = 312.9	Δ = 2272.6
f^4, $^5I_4$		
Component		
Γ	− 1.4	− 139.2
Φ	+ 35.7	+ 151.2
Δ	+ 106.6	+ 865.9
Π	− 56.9	− 306.0
Σ^+	− 168.0	− 1143.8
	Δ = 274.6	Δ = 2009.7
f^8, $^7F_6$		
Component		
Θ	− 226.2	
H	+ 27.9	
Γ	+ 95.3	
Φ	+ 77.5	
Δ	+ 34.4	
Π	− 1.5	
Σ^+	− 15.0	
	Δ = 321.5	

All energies in cm^{-1}.
[a]) Using $[Ce(Cot)_2]^-$ parameters of Table 1.
[b]) Using $U(Cot)_2$ parameters of Table 1.

As noted above the $4f^4$ complex, $[Pm(Cot)_2]^-$, has not yet been prepared, but a brief comment upon its predicted magnetic properties is in order since such measurements would provide a searching test of the present theoretical model. Thus the parameters of Table 1 indicate that the $\Sigma^+(M_J = 0)$ component of the $^5I_4$ ground level should lie lowest by a considerable margin (see Table 6). Consequently one would expect that for this system, as T → 0, the average moment should approach zero, leaving only the temperature independent paramagnetism due to second order Zeeman contributions, whilst at higher temperatures $\langle\mu\rangle$ should rise steeply before levelling off towards the free ion value. A similar conclusion follows also for the $5f^4$ complexes – $Pu(Cot)_2$ and $[Np(Cot)_2]^-$ – which appears qualitatively to be fulfilled, but there are here certain complicating feature (vide infra).

For the $4f^8$ system, $[Tb(Cot)_2]^-$, the parameters of Table 1 show (Table 7) that of the seven M_J components of the $^7F_6$ ground state the $\Theta(M_J = \pm 6)$ level is predicted by a very substantial amount to lie lowest, with a total splitting, Δ, of some $320\,cm^{-1}$. However, although the $\mu_\parallel$ and $\mu_\perp$ values thus calculated are markedly anisotropic (see Table 7) the average moment varies relatively little between 0 K and room temperature, so that little significance attaches to the reported moment value of 9.86 B.M. (*21*). Once again however experimental measurements of the magnetic anisotropy would constitute a valuable test of the theoretical approach. For the f^8 system the matrix elements of the Zeeman operators and the derived expressions for $\mu_\parallel$ and $\mu_\perp$ are given in the Appendix of reference (*26*).

For the 5f complexes the situation as regards experimental results is much better and magnetic moments have been determined over various temperature ranges for a large number of the actinocenes and their substituted derivatives. For the $An^{IV}(Cot)_2$ compounds these include $U(Cot)_2$, $U(EtCot)_2$, $U(n\text{-}BuCot)_2$, $U(Me_4Cot)_2$, $U(C_6H_5Cot)_2$, and $U((C_6H_5)_4Cot)_2$, together with the parent compounds and the $(EtCot)_2$ and $(n\text{-}BuCot)_2$ compunds of Np and Pu, and the $(Me_4Cot)_2$ derivative of Np, whilst of the $[An^{III}(Cot)_2]^-$ anions the Np, Pu, and Am complexes have so far been investigated. As noted before there are unhappily no magnetic studies of the

Table 7. Calculated Temperature Dependence of Magnetic Moment for a $4f^8$ System, in C^*_∞ Symmetry.

kT	$\mu_\parallel$	$\mu_\perp$	$\langle\mu\rangle$
0	15.59	0.00	9.00
25	15.59	2.01	9.15
50	15.38	3.37	9.29
100	13.76	6.46	9.53
150	12.41	7.89	9.63
200	11.64	8.52	9.67
300	11.09	9.04	9.77

All values assume k' = 1 ; free ion limit = 9.72 B.M.

5f[1] complex, $Pa(Cot)_2$, but uranocene has now been extensively studied, and a fairly consistent picture of its magnetic behaviour between 1.25 and 300 K is currently emerging. The magnetic data for each $An(Cot)_2$ system and its substituted analogues will therefore be considered, whereafter the more limited results for the $[An(Cot)_2]^-$ system will be surveyed.

For uranocene the first magnetic moment measurements, due to *Karraker*, *Stone*, *Jones*, and *Edelstein* (*13*), date from 1970, and a $\langle\mu\rangle$ value of 2.43 B.M., following

Table 8. Magnetic Properties of $An^{IV}(Cot)_2$ and $[An^{III}(Cot)_2]^-$ Systems.

Compound	Temperature Range	μ_{eff} (B.M.)	Θ °K	Other Information
$U(Cot)_2$ [a]	4.2 – 45	2.43	9.56	
$U(Cot)_2$ [b]	10 – 300	2.6	–	
$U(Cot)_2$ [c]	4 – 10	3.33	9.4	Inflection in χ^{-1} vs T
	10 – 42	2.3	0.9	plot at ca. 10 K
$U(Cot)_2$ [d]	1.25 – 6	–	–	TIP, $\chi = 75000 \times 10^{-6}$ cgsu
	12 – 70	2.40	2.9	Inflection in χ^{-1} vs T
	180 – 300	2.6	– 3.0	plot ca. 80–150 K
$U(EtCot)_2$ [c]	3 – 8	2.86	14.9	Inflection in χ^{-1} vs T
	10 – 27	1.9	0.4	plot at ca. 9 K
$U(n\text{-}BuCot)_2$ [c]	3 – 10	2.85	5.8	Inflection in χ^{-1} vs T
	10 – 50	2.3	2.6	plot at 10 K
$U(C_6H_5Cot)_2$ [e]	1.5 – 8	–	–	TIP, $\chi = 36000 \times 10^{-6}$ cgsu
	14 – 100	2.65	12.2	
$U(Me_4Cot)_2$ [e]	1.9 – 73.7	2.2	11.3	
$U((C_6H_5)_4Cot)_2$ [e]	4.2 – 100	2.5	6.7	
$Np(Cot)_2$ [a]	4.2 – 45	1.81	0.88	
$Np(Cot)_2$ [c]	3 – 60	1.64	0	
$Np(EtCot)_2$ [c]	3 – 50	1.82	1.74	
$Np(n\text{-}BuCot)_2$ [c]	3 – 50	1.71	0	
$Np(Me_4Cot)_2$ [c]	3 – 20	1.9	0.62	
$Pu(Cot)_2$ [a,c]	4.2 – 45	Diamagnetic		Anomalous increase in dia-
$Pu(EtCot)_2$ [c]	4.2 – 45	Diamagnetic		magnetism with decrease in
$Pu(n\text{-}BuCot)_2$ [c]	4.2 – 45	Diamagnetic		temperature
$[Np(Cot)_2]^-$ [f]	5.5 – 19	1.04	6.4	Inflection in χ^{-1} vs T
	19 – 80	1.39	26	plot at ca. 19 K
	2.2 – 5.5	–	–	TIP, $\chi = 11000 \times 10^{-6}$ cgsu
$[Pu(Cot)_2]^-$ [f]	2.2 – 80	1.25	3.7	
$[Am(Cot)_2]^-$ [g]	2.1 – 55	–	–	TIP, $\chi = 500 \times 10^{-6}$ cgsu (15–55 K)

[a]) Reference (*13*) ; [b]) Reference (*14*); [c]) Reference (*18*); [d]) Reference (*15*);
[e]) Reference (*19*) ; [f]) Reference (*63*) ; [g]) Reference (*64*).

the Curie-Weiss law ($\chi = C/(T + \theta)$), was reported for the range 4.2–45 K, with $\theta = 9.56°$. Data were also recorded for $Np(Cot)_2$ and $Pu(Cot)_2$ and an attempt was made to analyse the results in terms of a pseudo-axial crystal field model. However, due to the adoption of a point charge approach, it was concluded that the second order terms in V_{ax} would be very much greater than the fourth and sixth order contributions, and when these latter are neglected the purely electrostatic model leads to the $\Gamma(J_z = \pm 4)$ component of $^3H_4$ lying lowest, taking $A_2^0 \langle r^2 \rangle$ to be positive. By the same reasoning a $H^*(J_z = \pm 9/2)$ ground component was predicted for the $5f^3$, $^4I_{9/2}$, $Np(Cot)_2$, with $\Sigma^+(J_z = 0)$ lying lowest for the f^4, $^5I_4$, $Pu(Cot)_2$; on this basis therefore the model was at least qualitatively satisfactory since it correctly predicted the f^4 system to be diamagnetic (vide infra), and although the moments for $U(Cot)_2$ and $Np(Cot)_2$ predicted by the $g\,|M_J|$ values (3.20 and 3.27 B.M. respectively) were appreciably greater than the experimental results, the former could be brought into agreement if an orbital reduction factor, k′, of about 0.8 were introduced into the value of g, which would then become $g - (1 - k')(2 - g)$. It was however realised that the value of k′ required to achieve such agreement was rather low for actinide complexes, being nearer to those found for d-block complexes.

Subsequently however more detailed studies by *Karraker* (*18*) of the magnetic susceptibilities of $U(Cot)_2$ and its (EtCot) and (n-BuCot) derivatives revealed that the situation was somewhat more complex than had been supposed. Thus all three compounds showed a break in the $1/\chi$ vs T plot between 8 and 10 K, with the slope below that temperature corresponding to an appreciably larger magnetic moment than that deduced from the upper part of the temperature range. Thus for $U(Cot)_2$ between 10 and 42 K $\langle \mu \rangle$ is found as 2.3 B.M., with θ only 0.9 K, whereas from 4 to 10 K the corresponding values were 3.33 B.M. and 9.4 K, similar behaviour being also observed for the (EtCot) and (n-BuCot) complexes (see Table 8). From these findings *Karraker* (*18*) showed that the moment derived from the lower part of the temperature range was in good accordance with that predicted as $T \to 0$ for a $\Gamma(J_z = \pm 4)$ ground component (3.20 B.M.), whilst that obtained from the higher temperature range was close to that expected for a $\Phi(J_z = \pm 3)$ component (2.40 B.M.), leading to the conclusion that a $J_z = \pm 4$ level lay lowest, with the $J_z = \pm 3$ component slightly higher. The simple model of *Karraker et al.* (*13*) would though have predicted the $\Delta(J_z = \pm 2)$ level at about 20 K, for which no evidence was forthcoming, and it was therefore concluded (*18*) that the discontinuity was more likely to be due to the splitting of the U^{IV} $^3H_4$ ground state by both pseudo-axial and lower symmetry fields in the crystal. In support of this latter view it was pointed out that the low temperature discontinuity was not observed for any of the Np and Pu species, which significantly have respectively either a Kramer's doublet ground level or a $\Sigma^+(J_z = 0)$ lowest component, which could not be split by any field.

Nevertheless, just prior to the appearance of *Karraker's* (*18*) more detailed measurements, *Hayes* and *Edelstein* (*27*), using a Wolfsberg-Helmholz type molecular orbital approach, had concluded that the earlier (*13*) ideas of *Karraker et al.*, based on the weak crystal field model, required some modification. Moreover, despite numerical errors in the group overlap integrals there used (*27*), the results of the Hayes and

Edelstein treatment are in essential agreement with those of *Warren* (*26*), both predicting the $\Phi(J_z = \pm 3)$ level to be the ground component for uranocene. The values of $\langle\mu\rangle$ for the various uranium Cot complexes, listed in Table 8, are for the higher (and major) part of the temperature range clearly much more consistent with a $J_z = \pm 3$ lowest level whilst the low temperature discontinuity, although not fully understood, may reasonably be attributed to a lower symmetry ligand field contribution.

However, the picture so far built up has been challenged by the conclusions of *Amberger*, *Fischer*, and *Kanellakopulos* (*15*), who measured the susceptibility of uranocene from 298 K down to 1.25 K, and especially by their finding that between 1.25 and 6 K the susceptibility showed virtually temperature independent paramagnetism. This temperature independent susceptibility was very large (75000×10^{-6} cgsu) and since the μ^2 vs T plot between 1.25 and 6 K could be extrapolated through the origin it was concluded that the ground component was magnetically inactive (a $J_z = 0$ level) and that the off-diagonal matrix elements of the Zeeman operators within the ground state manifold were large, and the energy separations small. Between 12 and 70 K Curie-Weiss behaviour was found ($\theta = 2.9°$), with a maximum μ_{eff} of about 2.40 B.M., but above 80 K a steplike rise in the μ^2 vs T plot was observed, followed by a quasi-horizontal region between 180 K and room temperature, corresponding to a moment of some 2.6 B.M., with $\theta = -3.0°$. The data at the lower temperatures agreed well with those of *Karraker* (*18*), but *Amberger et al.* (*15*) favoured the conclusion that the ground level was $J_z = 0$, and the magnetic moment variation with temperature could be reproduced in the assumption that the successive excited states were $J_z = \pm 1, \pm 2$, and ± 3, at energies of 17, ca. 200, and ca. 3000 cm^{-1} respectively. The possibility could not however be ruled out that an additional ligand field of C_2 symmetry was operating, thereby splitting the ground level by about 20 cm^{-1}, and on this basis the lowest component could have been either $J_z = \pm 3$ or $J_z = \pm 4$.

Furthermore, assuming a $J_z = 0$ ground component, it was found (*15*) that to reproduce the splitting pattern deduced from the magnetic data it was necessary that all the B_0^n parameters should be negative with $B_0^2 = -11220$, $B_0^4 = -11408$, and $B_0^6 = -11680$ cm^{-1}. (Note that the conventions adopted by *Amberger et al.* for their splitting parameters differ from those of *Wybourne* (*35*).) These values are substantially greater than those commonly found for M^{IV} systems of the 5f series (*35*), and also appreciably exceed the definitive B_0^n values recently found (*59, 60, 61*) for the $5f^1$ $PaHal_6^{2-}$ anions. Moreover, the B_0^n parameters of *Amberger et al.* (*15*) are so large that very substantial J mixing due to V_{ax} is predicted, which would vitiate any attempt to calculate splitting patterns from diagonal V_{ax} matrix elements alone, and they also exceeded substantially the optimum B_0^n quantities which could be obtained (*15*) by a Wolfsberg-Helmholz molecular orbital calculation ($B_0^2 = -4792$, $B_0^6 = -1184$, $B_0^6 = -880$ cm^{-1}). Thus since even these latter parameters required the $H_M(5f)$ term to be so low (– 13.1 eV) as to be inherently inconsistent with the one-electron ordering scheme, it is clear that the question of which constitutes the lowest J_z level in uranocene cannot be answered unequivocably in favour of the magnetically inactive component.

Very recently however additional experimental evidence has been forthcoming which favours unambiguously the *Hayes* and *Edelstein* (*27*) scheme. Thus, *Edelstein, Streitwieser, Morrell,* and *Walker* (*19*) have measured the low temperature susceptibilities of $U(Me_4Cot)_2$, $U((C_6H_5)_4Cot)_2$, and $U(C_6H_5Cot)_2$, whose gross electronic structures may be expected, on good experimental grounds (*16, 17*), to be basically similar. Nevertheless, measurements between 1.5 and 100 K show that of the three systems only $U(C_6H_5Cot)_2$ resembles uranocene itself by showing Curie-Weiss behaviour between 14 and 100 K and temperature independent paramagnetism below 8 K: for $U(Me_4Cot)_2$ and $U((C_6H_5)_4Cot)_2$ on the other hand Curie-Weiss behaviour is found down to the lowest temperatures for which readings were taken (1.9 and 4.2 K respectively), effective moments of 2.2 and 2.5 B.M. being found (see Table 8). The latter observations are clearly inconsistent with a $J_z = 0$ ground level, and in addition the level ordering of *Amberger et al.* (*15*), which gave a total $^3H_4$ splitting of 6800 cm^{-1}, was found to predict $\Gamma^*(J_z = \pm 7/2)$ and $\Gamma(J_z = \pm 4)$ ground states for the f^3 and f^4 $Np(Cot)_2$ and $Pu(Cot)_2$ respectively. Experimentally $Pu(Cot)_2$ is found to be diamagnetic, whilst for $Np(Cot)_2$ the observed moment (ca. 1.8 B.M.) is quite consistent with the value (1.818 B.M.) calculated as $T \rightarrow 0$ on the *Hayes* and *Edelstein* (*27*) model for a $\Phi^*(J_z = \pm 5/2)$ ground component, but much too small for the value of 2.52 B.M. similarly predicted for a $\Gamma^*(J_z = \pm 7/2)$ ground level.

On balance therefore the Hayes and Edelstein model, with a $J_z = \pm 3$ ground state split by a small C_2 field must be preferred for uranocene, although the possibility of a similarly split $J_z = \pm 4$ ground level cannot be ruled out if an orbital reduction factor of about 0.8 is realistic. However, as pointed out by *Amberger et al.* (*15*), there is at least one piece of experimental evidence which if available could resolve the problem at once. Thus, for $J_z = 0$ lying lowest one expects $\chi_{\parallel} < \chi_{\perp}$, whereas for $J_z = \pm 3$ the opposite result $\chi_{\parallel} > \chi_{\perp}$, should be found. For the highly air-sensitive uranocene itself the measurement of the anisotropic paramagnetic susceptibilities would would be extremely difficult, but for the remarkably stable $U((C_6H_5)_4Cot)_2$, whose structure is also known (*62*), such measurements should be much easier, and would be invaluable in resolving this question.

For $Np(Cot)_2$ and its derivatives it is noteworthy that in all cases the Curie-Weiss plot may be extrapolated to pass very close to the origin, with extremely small θ values. Thus one may eliminate the possibility of there being low lying excited states closely juxtaposed to the ground level, as in uranocene, and the additional C_2 ligand field there postulated is clearly not operative in the $5f^3$ system. The values of the moments obtained for the low temperature range are closely bracketed around the limiting ($T \rightarrow 0$) value of 1.818 B.M. expected on the basis of the $\Phi^*(J_z = \pm 5/2)$ ground component, thus supporting the predictions based on the Hayes-Edelstein model (*26, 27*). From the B_0^n parameters calculated for $U(Cot)_2$ *Warren* (*26*) predicted only a small increase in the magnetic moment of $Np(Cot)_2$ over the temperature range studied, so that the experimental data are in good agreement with the theoretical expectations, requiring an orbital reduction factor, k′, of about 0.94. However, if the temperature range is extended up to around 300 K an increase of some 0.9 B.M. is predicted, and although this is entirely dependent on the calculated

B_0^n used, it would nevertheless be useful to have magnetic susceptibility data extending over the wider range. Thus, for $U(Cot)_2$ the data between 1.25 and 300 K substantially bear out the prediction of only a small increase in $\langle \mu \rangle$.

As noted before the values of k′ deduced from the magnetic data should be treated with caution, but it does appear that the larger k′ value, 0.97, required to accommodate the data for $U(Cot)_2$ is at least consistent with the simple molecular orbital picture. Thus it was shown by *Warren* (*26*) on this basis that the lowest lying $\Phi(J_z = \pm 3)$, and $\Phi^*(J_z = \pm 5/2)$ components of the $5f^2$ and $5f^3$ systems $U(Cot)_2$ and $Np(Cot)_2$ respectively should contain overall about 2% and 5% of ligand character, thus agreeing with the finding $k'(Np(Cot)_2) < k'(U(Cot)_2$. In fact nmr data suggest that the extent of covalency is slightly *less* for $Np(Cot)_2$ than for $U(Cot)_2$, but the larger proportion of ligand character calculated for the ground component of the Np compound arises from the appreciably greater amount of δ character in the $5f^3\,\Phi^*$ level as opposed to the $5f^2\,\Phi$ level. Finally it is of interest that the k′ value deduced for $U(Cot)_2$ also supports the $\Phi(J_z = \pm 3)$ ground component, whereas the $\Gamma(J_z = \pm 4)$ level earlier proposed by *Karraker et al.* (*13*) would necessitate a much smaller k′.

Of the plutonium compounds $Pu(Cot)_2$, $Pu(EtCot)_2$, and $Pu(n\text{-}BuCot)_2$ have all been found to be diamagnetic (*13, 18*) but, although the room temperature susceptibility of $Pu(Cot)_2(-218 \times 10^{-6}$ cgsu) was in good agreement with that calculated from the susceptibilities of its constituents (-187×10^{-6} cgsu), at lower temperatures the diamagnetism showed a large and unusual increase with decreasing temperature, reaching a value of -5000×10^6 cgsu at 4.2 K. This phenomenen totally obscures any trend in the magnetic moment along the lines predicted earlier, but the room temperature susceptibility is at least qualitatively consistent with the expectation that the $\Sigma^+(J_z = 0)$ level of $^5I_4$ should lie lowest, and also suggests that its separation from the higher, magnetically active, components must be appreciable. Attempts by *Craig* and *Stiles* (*65*) to obtain a theoretical explanation of the anomalous diamagnetism were unsuccessful, but *Karraker* (*66*) has informed the author that he is inclined to attribute the experimental results to unavoidable imprecision in the measurements and the presence of small but significant amounts of impurities, due to the highly reactive nature of $Pu(Cot)_2$, and doubts therefore whether detailed analysis of the data is justified.

For the $[An^{III}(Cot)_2]^-$ anions magnetic moment measurements have been carried out for the Np, Pu, and Am derivatives (*63, 64*), all in the form of the 2 THF solvates. Thus X-ray diffraction data show that $KNp(Cot)_2 \cdot 2\,THF$ and its Pu and Am analogues are all iso-structural, and similar results for the related diglyme solvates indicate that it is extremely probably that these are iso-structural with the salt $KCe(Cot)_2 \cdot$ diglyme, which is known (*23*) to have a sandwich structure with D_{8d} symmetry about the Ce atom. Consequently it is reasonable to assume that the pseudo-axial ligand field model is applicable to these $[An^{III}(Cot)_2]^-$ systems, and to interpret the magnetic data in these terms.

For $KNp(Cot)_2 \cdot 2\,THF$ *Karraker* and *Stone* (*63*) found a rather complex temperature dependence of the susceptibility, in which three distinct regions could be discerned. Thus below 5.5 K temperature independent paramagnetism was observed, but between 5.5 K and 19 K and above 19 K the χ^{-1} vs T plot approximated to two different Curie-Weiss lines (see Table 8), although some degree of curvature is evident, especially towards higher temperatures. Qualitatively this behaviour is very much what might be anticipated for a $5f^4$ system; thus the temperature independent paramagnetism is obviously consistent with a $\Sigma^+(J_z = 0)$ ground component, but its magnitude (11000×10^{-6} cgsu) and its rapid supersession by a temperature dependent region are both indicative of a first excited state lying only slightly above the ground level. Thus *Karraker* and *Stone* (*63*) analysed the data in terms of a $J_z = 0$ ground component with the $J_z = \pm 1$ and $J_z = \pm 2$ levels lying respectively only about 5 and 15 cm^{-1} higher: the calculations of *Hayes* and *Edelstein* (*27*) and of *Warren* (*26*) both agree in predicting a $\Pi(J_z = \pm 1)$ level as the first excited state in $5f^4$ systems, and although neither suggests $J_z = \pm 2(\Delta)$ as the next component it must be remembered that both calculations strictly relate only to An^{IV} systems. It seems likely however that any attempt to obtain a detailed fit to the magnetic data, using the ground state splittings, should take into account the substantial degree of impurity in the weak field $^5I_4$ ground state, due to the intermediate coupling, not to mention the possible effects of J mixing by the ligand field.

It is however noticeable that the splittings which have been deduced are very much smaller than those which have been attributed to the $An^{IV}(Cot)_2$ systems. Nevertheless there is a substantial body of experimental evidence (see for example *Wybourne* (*35*)) to support the contention that the B_0^n parameters are markedly smaller for An^{III} than for An^{IV} species, and the small covalent interaction in the former may be attributed to the greater shielding of the 5f orbitals in the larger An^{3+} ions. A similar arguement would lead to the expectation that the splitting parameters for An^V systems should be larger than those for An^{IV} complexes, and confirmation of this is provided by a study of the electronic spectra of U^V and Pa^{IV} hexahalo anions recently made by *Brown*, *Edelstein*, and *Whittaker* (*59, 60, 61*). In addition there is also direct evidence from Mössbauer spectroscopy indicating that the Np^{IV} Cot system are markedly more covalent then the Np^{III} compounds (vide infra).

For the $5f^5$ system, $KPu(Cot)_2 \cdot 2$ THF the magnetic susceptibility followed the Curie-Weiss law over the whole range studied, leading to an effective moment of 1.25 B.M. This is close to the value appropriate for the free ion Pu^{3+} when intermediate coupling is taken into account, thus suggesting that all the J_z components of the $^6H_{5/2}$ ground state may lie close together. However, it may equally well be that this agreement is fortuitous and it was concluded (*63*) that more experimental and theoretical work would be needed before the splitting patterns in these systems could be fully understood.

Very recently a magnetic study of the $5f^6$ complex, $KAm(Cot)_2 \cdot 2$ THF has been reported by *Karraker* (*64*), temperature independent paramagnetism being found between 15 and 55 K (below 15 K measurements were hindered by the presence of trace amounts of paramagnetic impurities giving temperature dependent susceptibilities). The small magnitude of the temperature independent susceptibility (500×10^{-6} cgsu), in contrast to the $5f^2$ and $5f^4$ cases, and its persistence to relatively high temperatures is fully consistent with the expected $^7F_0$ ground state since with

J = 0 this can only contain the magnetically inactive $J_z = 0$ component, and the temperature independent susceptibility must arise by second order Zeeman interactions involving higher J states (e.g. the anticipated first excited state, $^7F_1$).

As noted before the available electronic spectra data for the f-orbital biscyclooctatetraene complexes are lamentably few. Thus for the $[Ln^{III}(Cot)_2]^-$ complexes *Hodgson et al.* (*21*) have reported some fragmentary spectra for the near infra-red region (mostly between 5 and 15 kK) for the Pr, Nd, and Sm species in THF solution, which show the peak groups to occur in positions roughly comparable with those found for f–f transitions in the free Ln^{3+} cations, but some discrepancies are apparent, especially in the Pr complex. The data were not considered sufficient to attempt any systematic fitting, but the splittings observed tentatively suggest that the B_q^k parameters calculated (*26*) (Table 1) may be slightly too small. The peak positions and extinction coefficients of stronger bands in the visible region were also reported, but although these may represent either charge-transfer or f–d transitions there was insufficient evidence upon which to base an assignment.

In the $An^{IV}(Cot)_2$ series measurements of the lower energy region of the electronic spectra have been reported by *Karraker et al.* (*13*) for $Np(Cot)_2$ and $Pu(Cot)_2$ in toluene solutions, and a number of weak ($\epsilon \sim 10$–100) bands, assigned as f–f transitions, observed in the ranges 7–14 and 9.6–15 kK respectively. For $U(Cot)_2$ Schatz and Mowery are reported (*25*) as having found similar bands in the 10–12 kK region, but in no case has any attempt been made to analyse the results in terms of the appropriate B_0^n parameters. Nevertheless, the observation of such weak transitions, attributable to f–f transitions, provides some useful additional evidence to support the postulated bonding arrangement as represented in Fig. 1.

For the $[An^{III}(Cot)_2]^-$ complexes of Np and Pu attempts to measure the absorption spectra were frustrated by the presence of $An^{IV}(Cot)_2$ impurities (*63*), but for $[Am(Cot)_2]^-$ (*64*) the spectrum observed in THF solution between 8 and 22 kK was very similar to that previously found for Am^{3+} halides, thus suggesting that there is little covalent mixing of the metal f- and ligand orbitals, as was also found to be the case for the Np and Pu species (*63*).

More definitive information has however been forthcoming from a study of the charge-transfer spectra of the actinocenes and their alkyl substituted derivatives. Thus uranocene shows an extremely characteristic absorption in the visible region of the spectrum between about 14.5 and 17 kK, consisting of a sharp peak at 16.27 kK ($\epsilon = 1850$), with weaker subsidiary peaks towards lower energies at 15.6 (890), 15.16 (600), and 14.71 kK (350) respectively (*17*). The extinction coefficients strongly suggest that these are not f–f transitions, and this is borne out by the fact that alkyl or phenyl substitution brings about significant shifts of all the peaks to lower energies. Thus *Streitwieser* and *Harmon* (*17*) found rather small shifts for $U(EtCot)_2$, the four peaks being moved by only 0.08, 0.03, 0.11, and 0.24 kK respectively, whilst the corresponding values for $U(n\text{-}BuCot)_2$ were 0.16, 0.19, 0.22, and 0.26 kK: this is perhaps not too surprising for a single alkyl group in an eight-membered ring, but phenyl substitution produced shifts of 0.23, 0.22, 0.22, and 0.29 kK, and the dramatically large shifts found for $U(Me_4Cot)_2$ – 0.85, 1.04, 0.92, and 0.98 kK –

left no doubt that this band system arose from a ligand to metal charge-transfer excitation. The most reasonable assignment of this band is therefore as a transition from the highest occupied ligand orbital of even symmetry (the dominantly ligand e_{2g} orbital) to an odd symmetry metal f-orbital; in D_{8h} symmetry a transition to either the $e_{1u}(l_z = \pm 1)$ or the $e_{3u}(l_z = \pm 3)$ level would be permitted, but *Streitwieser* and *Harmon* preferred the former assignment because of the better spatial overlap expected. It is also of interest that alkyl substitution would be expected to increase the energies of the ligand π-orbital combinations, thus increasing the bonding interaction of the $e_{2u} - f(l_z = \pm 2)$ orbitals, and this is in accordance with the markedly greater stability observed (*17*) for the substituted uranocenes.

For neptunocene and plutonocene very similar findings have been recorded by *Karraker et al.* (*13*), bands at 19.32, 17.85, 17.20, and 16.70 kK (ϵ = 1300, 600, 800, and 850 respectively) being found for $Np(Cot)_2$, whilst $Pu(Cot)_2$ showed peaks at 24.70, 24.35. 22.30, and 19.40 kK (ϵ = 670, 670, 400, and 300) with a stronger absorption (ϵ = 3100) at 32.9 kK. Once again alkyl substitution led to a definite shift of the band maxima to lower energies and *Karraker* (*18*) found with the neptunocene derivatives that these were of the order of 0.15 to 0.20 kK for the Et and n-Bu complexes but amounted to about 1 kK for the Me_4 compound. For the two Pu complexes studied, the Et and n-Bu derivatives, shifts of the order of 0.20 kK to lower energies were again observed, and the evidence strongly suggests that for both neptunocene and plutonocene these band groups are again due to ligand to metal charge-transfer transitions of the same type as found for uranocene. Incidentally, the increase in energy of the band groups from f^2 to f^3 to f^4 here observed would be consistent with a naive interpretation in terms of increasing electron repulsions with increased occupation of the f levels.

Confirmatory evidence of the participation of the f-orbitals in the bonding scheme has also recently been obtained (*52*) from a study of the photoelectron spectra of uranocene and thorocene. Formally $Th(Cot)_2$ is $5f^0$ and $u(Cot)_2$ $5f^2$ and the most obvious difference between the two spectra lies in the presence of an ionisation at 6.20 eV in $U(Cot)_2$ which is absent in the thorium complex, and therefore clearly attributable to an f-electron ionisation. In $U(Cot)_2$ this band has a half-width of about 0.3 eV, but shows no structure or splitting: this indeed is not surprising since although the weak field $5f^2$ ground state, $^3H_4$, should in principle lead to both the $^2F_{5/2}$ and $^2F_{7/2}$ states of $5f^1$, separated by $7\xi/2$ (here about 0.7 eV), *Cox et al.* (*67*) have shown, using the coefficients of fractional paranetage, that the intensity ratio should be 6 to 1 in favour of the $^2F_{5/2}$ state, and consequently the higher lying $^2F_{7/2}$ ionisation is probably submerged beneath the higher energy ligand ionisations. For $U(Cot)_2$ these occur at 6.90 and 7.85 eV, and were assigned by *Clark* and *Green* (*52*) to ionisations from the mainly ligand e_{2g} and e_{2u} levels respectively. The separation between these two bands is much greater than is customarily found in sandwich species with identical rings, thus suggesting that one of the e_2 levels (e_{2g} and e_{2u}) is significantly more involved in the bonding than the other and supporting the views of *Streitwieser et al.* (*25*) concerning the great importance of the e_{2u} ligand – f-orbital interaction. In $U(Cot)_2$ the next set of bands is found at 9.94 and 10.24 eV (shoulder at 10.48 eV) and these are attributed to e_{1g} and e_{1u} ionisations. Significantly the separation between these two bands is much smaller than for the e_2 pair, thus bearing out the supposed greater participation of the e_{2u} level in the bonding. It is noteworthy also that the ionisations at 7.62 eV (e_2) and 10.51 eV (e_1) in CpTiCot (*68*), which were assigned (*69*) as originating from the Cot ring, correspond very well to the similar ionisations found and assigned in uranocene and thorocene.

A number of nmr studies have also been carried out on the actinocenes, of which those of *Edelstein et al.* (*14*) and of *Streitwieser et al.* (*16*) are the most significant. *Edelstein et al.* measured the temperature dependence of the proton resonance shifts in $U(Cot)_2$ and pointed out that for the $^3H_4$ ground state the orbital and spin angular momenta would be coupled anti-parallel so that a positive hyperfine coupling constant due to the Fermi contact term would result in an upfield proton resonance shift. A substantial positive shift, referenced against Cot^{2-}, was found over the range 198 to 353 K, and in order to obtain the Fermi contact term the dipolar contribution was calculated in the usual way (*70*), using the known crystal structure for $U(Cot)_2$ (*22*), and assuming $\chi_\parallel > \chi_\perp \sim 0$, as would be the case for a $J_z = \pm 4$ component lying lowest, with the energies of the higher lying components large with respect to kT. The calculation of the dipolar term involves however the assumption of a value for $\chi_\parallel - \chi_\perp$, but even proceeding as above the Fermi contact term was found to be substantially positive, and were $\chi_\parallel - \chi_\perp$ to be negative the contact term would become even larger. The values deduced for the proton hyperfine coupling constants do however depend upon the assumed J_z value for the lowest lying $^3H_4$ component, and for numerically smaller J_z values for the ground component these coupling constants would be larger. Nevertheless, the question of which J_z component lies lowest does not affect the conclusion that the Fermi contact term is substantially positive, and on this basis it was concluded that an appreciable covalent interaction between the metal and the ligands did occur, although it was not possible with certainty to apportion this between the σ and π frameworks of the rings. This latter problem was in fact tackled by *Streitwieser et al.* (*16*) who determined the chemical shifts for $U(Cot)_2$, $Np(Cot)_2$, and their $(Me_4Cot)_2$ derivatives. In principle it is possible to determine whether the spin density in the rings resides mainly in a σ or a π type orbital by comparison of the contact terms for the ring and methyl protons of the substituted systems, since for in-plane (σ) orbitals these shifts will be of the same sign, with the ring proton some three to five times larger than the methyl shifts, whereas for π orbitals of the rings the shifts should be of comparable magnitude but opposite sign. Proceeding on the same assumptions as made by *Edelstein et al.* (*14*), the latter result was obtained, on which basis it was concluded that the spin density resided in the ligand π orbitals, a charge transfer mechanism from filled ligand π-orbitals to vacant f-orbitals being proposed. It should nevertheless be noted that the observed chemical shifts were rather small for the methyl protons of both the substituted compounds. Thus, since the calculated dipolar terms are appreciably larger for the methyl protons than for the ring, the value assumed for $\chi_\parallel - \chi_\perp$ becomes of crucial importance. This quantity would thus become rather smaller for a $J_z = \pm 3$ ground level and would be negative for $J_z = 0$.

Consequently, although the conclusion that the spin density is located essentially in the π-framework of the ring is in accordance with the established experimental data, it is necessary to reserve final judgement upon the subject until values of the anisotropic susceptibilities are known experimentally. Similarly, the conclusion that the extent of covalent interaction in $U(Me_4Cot)_2$ is somewhat greater than in $Np(Me_4Cot)_2$ is not unequivocably established since it depends upon the approximate identity of the hyperfine coupling constants in the two species (as deduced from the calculated Fermi contact terms), which contain respectively two and three unpaired electrons.

For the Cot_2 derivatives of Np^{III} and Np^{IV} it has proved possible to obtain valuable information about the extent of metal-ligand interaction by the use of Mössbauer spectroscopy (*13, 63*). For Np^{3+} in $KNp(Cot)_2 \cdot 2$ THF the isomer shift (relative to $NpAl_2$) at 77 K) of + 3.92 cm/sec. is the largest positive shift yet found for Np(III) and slightly more positive than that for $NpCl_3$ and $Np(Cp)_3 \cdot 3$ THF (+ 3.54 and + 3.64 cm/sec. respectively), but the covalent contribution is small by comparison with $Np(Cot)_2$, whose isomer shift is + 1.94 cm/sec. compared with – 0.34 cm/sec. for $NpCl_4$ and + 0.72 cm/sec. for $NpCp_4$. The negligible covalency in $[Np(Cot)_2]^-$ together with the much larger effects in $Np(Cot)_2$ parallels the pattern found (*71*) for $NpCp_4$ and $NpCp_3 \cdot 3$ THF, and reinforces the conclusion that metal-ligand mixing is much more significant in An^{IV} than in An^{III} systems.

Finally, it is appropriate to comment briefly on the remaining uncertainty concerning the magnetic behaviour of uranocene at very low temperature. In this context it is perhapt significant that, using the ligand field formalism described herein, only uranocene ($5f^2$) of all the f^x ground states considered is predicted to have a lowest lying J_z component which is orbitally degenerate and capable of being split by descent in symmetry; consequently the possible participation of Jahn-Teller effects should be accorded serious consideration.

At present the state of knowledge regarding the f-orbital sandwich complexes and their mode of bonding is both fragmentary and overdependent upon conjecture. This current attempt at rationalisation is thus merely tolerably consistent with the available experimental evidence, which it is to be hoped will soon be usefully augmented. High upon the list of desirable experimental information must however be a study of the magnetic and electronic spectral properties of the $5f^1$ complex, $Pa(Cot)_2$, esr investigations of f^1 and f^3 systems, measurements of anisotropic susceptibilities, and further spectral and mcd studies of all the actinocenes.

Acknowledgements: – I am grateful to *D.W. Clack* of this Department for allowing me to use the results of reference (*43*) prior to publication, and for many useful discussions, and I am similarly indebted to *N. Edelstein* with respect to reference (*19*). It should also like to thank *D.P. Craig, R.G. Hayes, D.G. Karraker*, and *R.L. Mowery* for helpful private communications.

References

1. *Coulson, C.A., Lester, G.R.:* J. Chem. Soc., 3650 (1956).
2. *Moffitt, W.:* In: *Reynolds, L.T., Wilkinson, G.:* J. Inorg. and Nucl. Chem., **2**, 246 (1956).
3. *Kettle, S.F.A., Pioli, A.J.P.:* J. Chem. Soc. A, 122 (1968).
4. *Jørgensen, C.K., Pappalardo, R., Schmidtke, H.H.:* J. Chem. Phys., **39**, 1422 (1963).
5. *Kettle, S.F.A., Smith, A.J.:* J. Chem. Soc. A, 688 (1967).
6. *Eisenstein, J.C.:* J. Chem. Phys., **25**, 142 (1956).
7. *Fischer, R.D.:* Theoret. Chim. Acta, 1, 418 (1963).
8. *Streitwieser, A., Müller-Westerhoff, U.:* J. Amer. Chem. Soc., **90**, 7364 (1968).
9. *Streitwieser, A., Yoshida, N.:* J. Amer. Chem. Soc., **91**, 7528 (1969).
10. *Goffart, J., Fuger, J., Gilbert, B., Kanellakopulos, B., Duyckaerts, G.:* Inorg. and Nucl. Chem. Letters, **8**, 403 (1972).
11. *Goffart, J., Fuger, J., Brown, D., Duyckaerts, G.:* Inorg. and Nucl. Chem. Letters, **10**, 413 (1974).
12. *Starks, D.F., Parsons, T.C., Streitwieser, A., Edelstein, N.:* Inorg. Chem., **13**, 1307 (1974).
13. *Karraker, D.G., Stone, J.A., Jones, E.R., Edelstein, N.:* J. Amer. Chem. Soc., **92**, 4841 (1970).
14. *Edelstein, N., La Mar, G.N., Mares, F., Streitwieser, A.:* Chem. Phys. Letters, **8**, 399 (1971).
15. *Amberger, H.D., Fischer, R.D., Kanellakopulos, B.:* Theoret. Chim. Acta, **37**, 105 (1975).
16. *Streitwieser, A., Dempf, D., La Mar, G.N., Karraker, D.G., Edelstein, N.:* J. Amer. Chem. Soc., **93**, 7343 (1971).
17. *Streitwieser, A., Harmon, C.A.:* Inorg, Chem., **12**, 1102 (1973).
18. *Karraker, D.G.:* Inorg. Chem., **12**, 1105 (1973).
19. *Edelstein, N., Streitwieser, A., Morrell, D.G., Walker, R.:* Inorg. Chem., **15**, 1397 (1976).
20. *Mares, F., Hodgson, K., Streitwieser, A.:* J. Organometal. Chem., **24**, C 68 (1970).
21. *Hodgson, K.O., Mares, F., Starks, D.F., Streitwieser, A.:* J. Amer. Chem. Soc., **95**, 8650 (1973).
22. *Avdeef, A., Raymond, K.N., Hodgson, K.O., Zalkin, A.:* Inorg. Chem., **11**, 1083 (1972).
23. *Hodgson, K.O., Raymond, K.N.:* Inorg. Chem., **11**, 3030 (1972).
24. *Baker, E.C., Halstead, G.W., Raymond, K.N.:* Structure and Bonding, **25**, 23 (1976).
25. *Streitwieser, A., Müller-Westerhoff, U., Sonnichsen, G., Mares, F., Morrell, D.G., Hodgson, K.O., Harmon, C.A.:* J. Amer. Chem. Soc., **95**, 8644 (1973).
26. *Warren, K.D.:* Inorg. Chem., **14**, 3095 (1975).
27. *Hayes, R.G., Edelstein, N.:* J. Amer. Chem. Soc., **94**, 8688 (1972).
28. *Clack, D.W., Warren, K.D.:* J. Organometal. Chem., **122**, C28 (1976).
29. *Scott, D.R., Matsen, F.A.:* J. Phys. Chem., **72**, 16 (1968).
30. *Warren, K.D.:* J. Phys. Chem., **77**, 1681 (1973).
31. *Jørgensen, C.K.:* Electrons in Atoms and Molecules. New York: Academic Press, 1962.
32. *Stevens, K.W.H.:* Proc. Phys. Soc. A, **65**, 213 (1952).
33. *Elliott, R.J., Stevens, K.W.H.:* Proc. Roy. Soc. A, **215**, 437 (1952).
34. *Abragam, A., Bleaney, B.:* Electron Paramagnetic Resonance of Transition Ions. London: Oxford University Press, 1970.
35. *Wybourne, B.G.:* Spectroscopic Properties of Rare Earths. New York: Interscience, 1965.
36. *Hutchings, M.T.:* Solid State Physics, **16**, 227 (1964).
37. *Hutchings, M.T., Ray, D.K.:* Proc. Phys. Soc., **81**, 663 (1963).
38. *Watson, R.E., Freeman, A.J.:* Phys. Rev., **127**, 2058 (1962).
39. *Lenander, C.J.:* Phys. Rev., **130**, 1033 (1963).
40. *Hillier, I.H., Canadine, R.M.:* Disc. Faraday Soc., **47**, 27 (1969).
41. *Schaffer, C.E., Jørgensen, C.K.:* Mol. Phys., **9**, 401 (1965).
42. *Gerloch, M., Slade, R.C.:* Ligand Field Parameters. London: Cambridge University Press, 1973.

43. *Clack, D.W., Smith, W.:* Rev. Roum. Chim., **20**, 1243 (1975).
44. *Mulliken, R.S., Rieke, C.A., Orloff, D., Orloff, H.:* J. Chem. Phys., **17**, 1248 (1949).
45. *Cotton, F.A.:* Chemical Aspects of Group Theory. New York: Interscience, 1963.
46. *Brink, D.M., Satchler, G.R.:* Angular Momentum. London: Oxford University Press, 1968.
47. *Jørgensen, C.K.:* J. de Physique, **26**, 825 (1965).
48. *Mowery, R.L.:* private communication.
49. *Hayes, R.G.:* private communication.
50. *Brewer, L.:* J. Opt. Soc. Amer., **61**, 1101, 1666 (1971).
51. *Radziemski, L.J., Steinhaus, D.W., Cowan, R.D., Blaise, J., Guelachvili, G., Osman, Z.B., Verges, J.:* J. Opt. Soc. Amer., **60**, 1556 (1970).
52. *Clark, J.P., Green, J.C.:* J. Organometal. Chem., **112**, C14 (1976).
53. *Clementi, E.:* Tables of Atomic Functions. San Jose: International Business Machines, 1965.
54. *Satten, R.A., Margolis, J.S.:* J. Chem. Phys., **32**, 573 (1960).
55. *Lang, R.J.:* Canad. J. Research, **A 14**, 127 (1936).
56. *Warren, K.D.:* Inorg. Chem., **13**, 1317 (1974).
57. *Schwartz, R.W., Hill, N.J.:* J. Chem. Soc. Faraday II, **70**, 124 (1974).
58. *Buchanan, R.A., Rast, H.E., Caspers, H.H.:* J. Chem. Phys., **44**, 4063 (1966).
59. *Edelstein, N., Brown, D., Whittaker, B.:* Inorg. Chem., **13**, 563 (1974).
60. *Brown, D., Whittaker, B., Edelstein, N.:* Inorg. Chem., **13**, 1805 (1974).
61. *Brown, D., Lidster, P., Whittaker, B., Edelstein, N.:* Inorg. Chem., **15**, 511 (1976).
62. *Streitwieser, A., Walker, R.:* J. Organometal. Chem., **97**, C41 (1975).
63. *Karraker, D.G., Stone, J.A.:* J. Amer. Chem. Soc., **96**, 6885 (1974).
64. *Karraker, D.G.:* 4th. International Transplutonium Element Symposium, Baden-Baden, Germany, 1975.
65. *Craig, D.P., Stiles, P.:* private communication.
66. *Karraker, D.G.:* private communication.
67. *Cox, P.A., Baer, Y., Jørgensen, C.K.:* Chem. Phys. Letters, **22**, 433 (1973).
68. *Green, J.C., Jackson, S.E., Higginson, B.R.:* J. Chem. Soc. Dalton, 304 (1974).
69. *Warren, K.D.:* Structure and Bonding, **27**, 45 (1976).
70. *McGarvey, B.R.:* J. Chem. Phys., **53**, 86 (1970).
71. *Goebel, J.B., Wilson, A.S.:* cited in reference (*63*).

Note added in proof:

Very recently P.N. Schatz has communicated to the author the results of a study (R.L. Mowery, Ph.D. thesis, University of Virginia, 1976) of the absorption spectrum and the MCD of sublimed films of uranocene between 14000 and 18000 cm^{-1} and in the temperature range 4–21 K.

The absorption spectrum shows a strong peak at 16200 cm^{-1} with weaker bands at 15500 cm^{-1} and at lower energies, but no temperature dependence is observed between 4 and 21 K. This therefore argues strongly against the model suggested by Amberger et al. since with an A_{1g} (Σ^+) ground level and an E_{1g} (Π) level at 17 cm^{-1} a significant temperature dependence would be anticipated. In the MCD however, which is positive throughout, the 16200 cm^{-1} band is not the strongest feature, and the spectrum is dominated by the 15500 cm^{-1} peak, with weaker peaks at lower energies. Below 16000 cm^{-1} the MCD shows an appreciable variation with temperature, although this is not proportional to 1/kT as for a normal C term, but the 16200 cm^{-1} peak is temperature independent over the range studied.

The temperature independence of the absorption spectrum suggests that any model requiring two or more D_{8h} states to lie effectively within the ground state may, with two exceptions, be rejected. However, ligand field calculations show that either a B_{1g}, B_{2g} (Γ) or an E_{3g} (Φ) ground state is probable, and the acceptable models require either a B_{1g}, B_{2g} ground level or an E_{3g} ground state, split by a low symmetry ligand field. In either case the results were found to be consistent with a separation of some 10–15 cm^{-1} between the two lowest lying components, but for the B_{1g}, B_{2g} case this situation may arise either by configuration interaction with levels of the same symmetry derived from ligand orbitals (the B_{1g} and B_{2g} levels are strictly degenerate in the crystal field model), or by a low symmetry ligand field contribution, whereas for an E_{3g} ground level only the latter mechanism may lift the degeneracy.

The MCD results do not however allow a definitive choice to be made between these alternatives. For an E_{3g} ground level the allowed electric dipole transitions are to B_{1u}, B_{2u}, and E_{2u} levels (x, y polarised) or to an E_{3u} level (z polarised), but for B_{1u} and B_{2u} states the MCD is predicted to be negative. For a B_{1g}, B_{2g} ground state the allowed transitions are to E_{3u} (x, y) or to B_{1u}, B_{2u} (z), but in either case the MCD was most reasonably interpreted by attributing the lower energy features (ca. 15500 cm^{-1}) to x, y polarised transitions, and the 16200 cm^{-1} band to a z polarised transition. Calculations showed however that this band should only acquire intensity in the MCD by magnetic dipole mixing between the ground and higher excited levels, thus accounting for its observed temperature independence. On this basis therefore the MCD, which contains C, C', and/or B contributions, may reasonably be understood, and although a final choice between a B_{1g}, B_{2g} (Γ) and an E_{3g} (Φ) level for the ground state is not possible, it was found that in either case the bands studied were most likely to be due to ligand → metal, $e_{2g} \rightarrow e_{3u}$ excitations.

Errata

K. D. Warren, Structure and Bonding, **27**, 45 (1976).
p. 62: Table 5, $^3\Pi$ matrix; the third diagonal entry should be $-12B + 4C$, instead of as shown.
p. 63: Table 5, $^3\Gamma$ matrix; the off-diagonal element should be $\sqrt{6}B$, instead of as shown.

I am grateful to V. Cerny for drawing my attention to these typographical errors.

The Two-Correlations Model, a Valence Model for Metallic Phases

K. Schubert

Max-Planck-Institut für Metallforschung (Institut für Werkstoffwissenschaften), Seestr. 75, Stuttgart, F.R. Germany

Table of Contents

Abstract

To build up a theory of metallic phase stability the conventional arguments for the discussion of crystal structures have to be completed by the concept of spatial correlation of electrons. The parameters of the spatial correlations may be analyzed from the ample empirical material of determined crystal structures by means of several evident rules. The surprising result is that for many crystal structures two correlations are essential for understanding the special features of the crystal structures. This two-correlations model makes possible an easy survey of metallic structures. Two examples of crystal structure-type families are considered, the Cu-family and the W-family.
In the Cu-family substitutional variants, simple shear variants, multiple shear variants, and displacive variants may be distinguished. In the W-family cubic substitutional variants, deformed substitutional variants, vacancy variants with weak deformations, variants with stronger deformations, and variants with row shear may be distinguished.
The essential feature of the present exposition is that the structural types are not only organized in a geometrical manner but that the arrangement of the structural types is put on an energetic base. This energetic base gives an answer to the fundamental question why certain phases are formed in certain mixtures.

I. Concepts for Assessing Alloy Phase Stability

One of the present problems of metal physics is the explanation of alloy phase stability. The number of different alloy phases occurring in industrial and scientific techniques is so great, and the requirements for the properties of these substances are so diversified, that only a better understanding of the alloy phase stability gives us the possibility to handle the overwhelming amount of information collected so far and to search for desired properties. However since a calculation of phase stability is not possible at present, the immediate problem, to find a qualitative energetic explanation of alloys, must be solved. A purely geometric discussion of structural types is only a part of the required explanation; it should be supplemented by arguments of energetic character *(structural arguments)*.

This energetic character becomes apparent on considering the thermodynamic description of alloys. We first need measures to define the *thermodynamic state*, for instance the extensives V, U, N (volume, internal energy, mol numbers). We then consider two systems in the states V_1, U_1, N_1 (=: M_1) and V_2, U_2, N_2,(=: M_2), combine them to form one system, wait a certain time and take care that ΣV, ΣU, ΣN remain constant during the formation of a new equilibrium state. We observe that the *entropy S becomes a maximum* $S(\Sigma_i M_i) \geqslant \Sigma_i S(M_i)$ as compared to the sum of entropies of the incipient states (entropy maximum law); this increase of S is caused by currents equalizing P, T, μ (pressure, temperature, chemopotentials) which may have been different at the beginning. Any constitutional function S_{VUN} may be measured with the aid of the Carnot entropy pump which allows fixed amounts of entropy to be pumped into a system together with a certain amount of energy at any value of V, U, N, so that S(U, V, N) is an experimentally known function. Another law of thermodynamics asserts $\partial S/\partial U =: \partial_U S > 0$; therefore the function S_{VUN} may be inverted to U_{VSN} and the entropy maximum law transforms into the *energy minimum law for V, S, N = const.* Instead of stability we can therefore also say *energetic preference.* In the case of the independent variables V, T, N or P, T, N we have to use the differential transforms $(1 - S\,\partial_S)\,U =: F_{VTN}$ or $(1 - V\,\partial_V - S\,\partial_S)\,U =: G_{PTN}$ and obtain the laws of minimum free energy. The free energy function per mol G′ for a mixture follows from the first degree homogenity of G in N as $G' = \mu_1(1 - N'_2) + \mu_2 N'_2$ where N'_2 is the mol fraction of the second component and $\mu_c = \mu^{id}_{cPTN'\downarrow 2\ldots} = \mu^{p}_{cPT} + RT\ln N'_c$ is the ideal chemopotential of the component *c* (μ^p_c being the chemopotential of the pure component *c*). The graph of this constitutional function shows that a continuous series of mixed crystals is described. In the more general case of a *regular solution* we have a quadratic excess Gibbs energy and may describe, for instance, a mixture with an interval of nonmiscibility. Similar mixtures have been calculated by *Kaufmann* and *Bernstein* 1970 with good success, but they could not derive the existence of intermetallic phases.

The problem of energetic preference of phases is therefore not described by macrophysics but is a *problem of binding*, that is, of the movement of electrons in the metal. For this movement a simple model exists: the *band model*. This model is useful for understanding transport phenomena in metals, but for the problem of binding it has not given helpful insights, since the one-electron-wave functions used in this theory are only an approximation to the truth and do not express in a simple manner the spatial correlations of the electrons, which are essential for assessing phase energy.

The crystallographers have instead used for a long time and with much success intuitive concepts, such as the *atomic radii* which are practically independent of binding, or such as the *electric charges* giving Madelung-type energies. An actual problem is whether these arguments are exhaustive, or amenable to and in need of supplementary arguments. Microscopic theory shows that they are not exhaustive. In order to see this, we consider all valence electrons of the crystal; we anticipate that at a certain temperature (or entropy) they obey a certain probability density D_{xp}, which is a function of the microscopic state x, p =: b (x = coordinates, p = momenta). We also know the energy function $H_{xp} = \Sigma_{in} p_{in}^2/2m + \Sigma_n U^P_{x\downarrow in} + \Sigma_{nn'} e^2/2|x_{in} - x_{in'}|$ (i = coordinate number, n = electron number, m = electron mass, e = electron charge, U^P = positional energy) so that the averaging law of probability theory gives us $U = \int_{xp} H_{xp} D_{xp}$ as the expectation value of the energy. This is the fundamental formula of classical statistics. Clearly the density function is a very broad concept, which depends on all the coordinates and all the momenta of the electrons of our system. We cannot use it, unless it is greatly simplified. Curiously enough the greatest simplification becomes possible in the quantum-mechanical version of the averaging law, which at first sight appears more complicated but which is, as we know, nearer to the reality than D_{xp}.

The function D_{xp} is composed of Diraç functions $D = \Sigma_b g_b \delta_b$ where $\Sigma_b g_b = 1$ $g_b \geqslant 0$ and the δs are forbidden by Planck's law of the finiteness of the quantum of action (which may be formulated as the indeterminacy principle). This may be corrected for by the so-called *Wigner transformation* which transforms the expectation value linear form to $U = \int_{xx'} H_{x'x} D_{xx'}$ where the coordinates x occur twice, independently, so that H and D become matrices. Since the Wigner transformation must lead to indeterminacy, it is closely related to a Fourier transformation and both matrices *H and D become hermitian* (Bopp, 1961). The dyads of hermitian matrices may be written $\psi_x \psi_{x'}^*$ and we see that their contribution u to the expected energy becomes $u = \int \psi^* H\psi$, therefore if we choose ψ_x as eigenvectors of H we see that we have in fact only *discrete possible states* in agreement with indeterminacy.

If we consider the energy we find that $\Sigma p^2/2m$ is the only non-diagonal contribution (remembering $p_{in} = h\,\partial_{x\downarrow in}/2\pi i$ in quantum mechanics). On the other side $\Sigma p^2/2m$ is not structure-sensitive, i.e., it does not contain interaction terms since the magnetic forces are negligible. We make therefore the widely used assumption that the energy difference of structures depends mainly on configurational energy, i.e., that the kinetic energy is approximately the same in all structures. This assumption removes the non-diagonal elements from the density matrix $D_{xx'}$ so that it becomes the configurational probability density D_x. The next simplification strongly reduces the number of variables. Since H contains at most two-electron contributions we may restrict to the so-called reduced density $D''_{x\downarrow i1 x\downarrow i2} = \int_{x\downarrow i3...} D_{x\downarrow i1...}$ which may be called the *spatial correlation of the electrons*. Because of the indistinguishability

$D_{x\downarrow i1, x\downarrow i2,\ldots} = D_{x\downarrow i2, x\downarrow i1\ldots}$ caused by indeterminacy, all possible reduced densities D'' are the same.

The concept of a spatial correlation is very rich in content, and its content is necessary and sufficient for energy assessment. It is useful to consider the spatial correlation in the approximation of the band model: from all the wave functions contained in the Fermi sphere a determinant has to be formed as, also the determinants of weakly excited states; a perturbation calculation has to be made with respect to $\Sigma e^2/|x_{in} - x_{in'}|$; the reduction to the spatial correlation must be made; the energy may be calculated. In fact the pertubation calculation has never been made for metals, while the spatial correlation was only calculated by *Wigner* and *Seitz* (1933) in the zero approximation. The result of their calculation is the well-known exchange hole, which may be represented in three-dimensional space if one electron is fixed in the coordinate origin. This may be done if the local space is homogenous i.e., has no

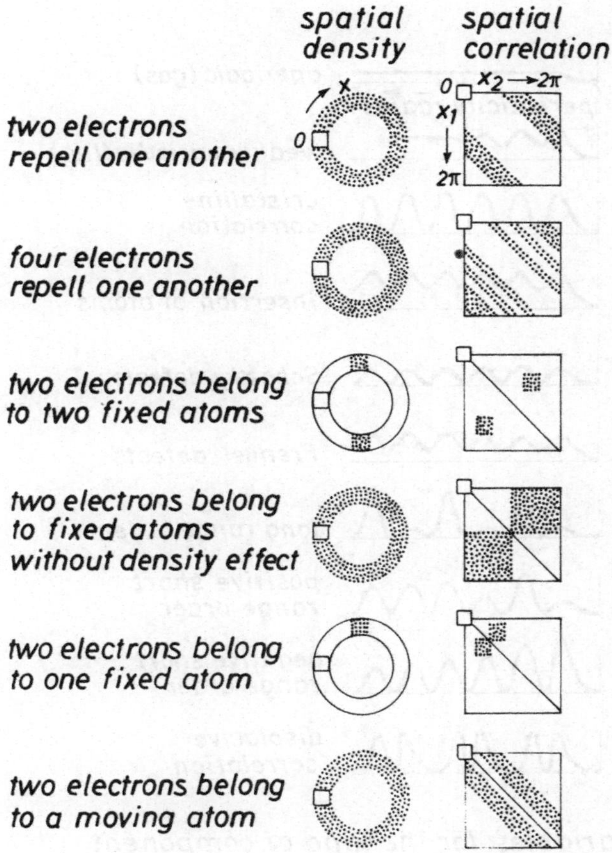

Fig. 1. Examples of spatial density and spatial correlation of particles in a finite circular space.

structure. For the electron spatial correlation in a non-idealized crystal, the three-dimensional representation is only possible by an averaging procedure. Examples for spatial correlations are given in Figure 1. Here the simplest example of a spatial correlation is considered where the coordinate of one particle is one-dimensional and finite, and therefore the space of the spatial correlation function D″ is two-dimensional and finite. In a three-dimensional crystal the space of D″ is six-dimensional and in principle infinite. Figure 1 gives several examples for the important fact that the spatial correlation is strongly "localized", while the spatial density which results from D″ by a reduction calculation is not at all localized. This shows that we cannot hope to investigate the essentials of binding in crystals by electron density measurements. In Figure 2 it is shown that the concept of a spatial correlation is quite familiar in chemistry and is equivalent for atomic systems to the concept of structure or radial distribution function. For the electron gas of a metal it is not favorable to speak of a structure, since it might be concluded that electron density effects are

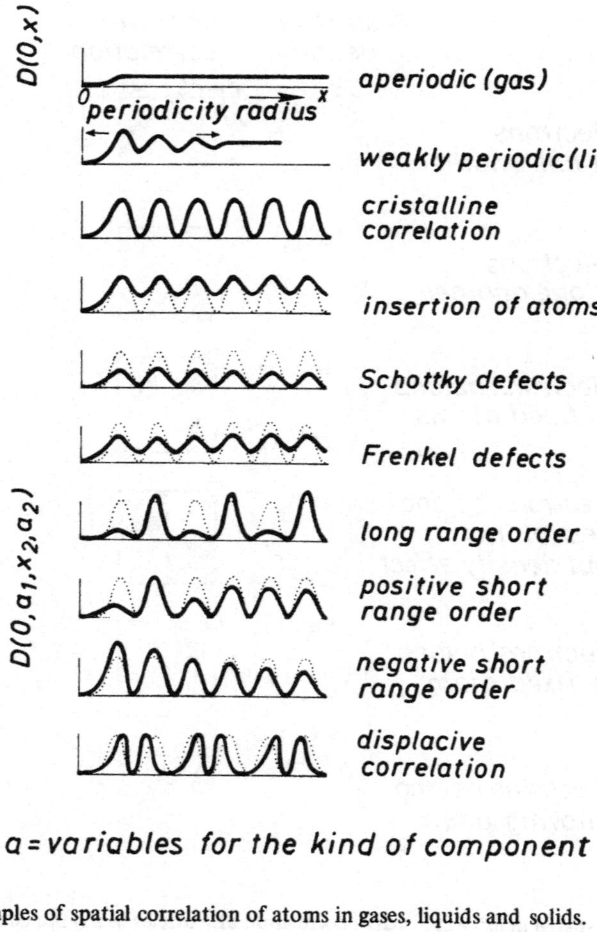

Fig. 2. Examples of spatial correlation of atoms in gases, liquids and solids.

implied, which is not the case, as Fig. 1 shows. From Fig. 2, Schottky defects, it becomes clear *that a special correlation may be partly occupied*. This may be understood by imagining the spatial correlation as a "momentaneous lattice" which contains vacancies.

From the spatial correlation we derive by integration over one coordinate triple a reduced density: the *spatial density of electrons*. The older intuitive energetic arguments are all statements on the spatial density: atomic radii describe the breadth of the density maxima and electric charges refer to the height of such maxima. Very few arguments using the spatial correlation have been used up to the present. The best known is the Wigner-Seitz exchange hole, but this stems from the antisymmetry of ψ-amplitudes belonging to electrons and not from their mutual repulsion. A further example is the Hylleraas wave function in the hydrogen molecule, which has also been extended to all He-like atoms and to small molecules (Hurley 1976). Since the experimental crystallographers have gathered so much knowledge on atomic radii and atomic charges in structures, the important task arises today to find from the empirical crystal structures indications for the spatial correlation of the valence electrons which are in first line responsible for the chemical interaction of atoms.

Two properties of the spatial correlation may instantly be anticipated: *lattice likeness,* to obtain a low energy of the considered valence electron gas, and *commensurability* to the crystal structure, to obtain a low interaction energy between electron gas and crystal structure. If $(a_{i1}, a_{i2}, a_{i3}) =: a_{is} =: a_s$ are the elementary vectors of a structure and $a'_{ic} =: a'_c$ the elementary vectors of a valence electron correlation then $a_{is} = \Sigma_c\, a'_{ic}\, K'_{cs}$ is a commensurability where $K'_{cs} = (K'_{11}, K'_{12}, K'_{13}; K'_{21}, K'_{22}, K'_{23}; K'_{31}, K'_{32}, K'_{33})$ is the *commensurability matrix.* The valence electron correlation and also the connected commensurability matrix are written with a prime in order to have a clear distinction from a lattice matrix of a crystal. In Fig. 3 a commensurability between crystal structures is written in the extended form and in shorter notations. The lattice matrix of Cu is given with tilde ~ in order to indicate

Full notation:

$$\begin{pmatrix} a_{11} & a_{12} & a_{13} \\ a_{21} & a_{22} & a_{23} \\ a_{31} & a_{32} & a_{33} \end{pmatrix} = \begin{pmatrix} \tilde a_{11} & \tilde a_{12} & \tilde a_{13} \\ \tilde a_{21} & \tilde a_{22} & \tilde a_{23} \\ \tilde a_{31} & \tilde a_{32} & \tilde a_{33} \end{pmatrix} \times \begin{pmatrix} 1 & 0 & 0 \\ 0 & 1 & 0 \\ 0 & 0 & 2 \end{pmatrix} = \begin{pmatrix} \tilde a_{11} & \tilde a_{12} & 2\tilde a_{13} \\ \tilde a_{21} & \tilde a_{22} & 2\tilde a_{23} \\ \tilde a_{31} & \tilde a_{32} & 2\tilde a_{33} \end{pmatrix}$$

in words:

cell of $TiAl_3$ = approximate cell of Cu × commensurability = result

short general notation: $a_{i,TiAl3} = \Sigma_{Cu}\, \tilde a_{i,Cu}\, K_{Cu,TiAl3}$

shorter and less general: $a_{TiAl3} = \tilde a_{Cu}(1,0,0;0,1,0;0,0,2)$

even shorter: $a = \tilde a_{Cu}(1;1;2)$

Fig. 3. Commensurability matrix in different notation.

that the a_{Cu} matrix must be deformed a little to get a whole numbered commensurability; it is seen that for convenience Cu or $TiAl_3$ is chosen as one matrix index. In electron correlation matrices the correlation type (two characters) is used as second matrix index, while the first matrix index "i" is often left out. A diagonal matrix may be written in abbreviated form $(K_{11}; K_{22}; K_{33})$, and instead of (K; K; K) we write (K).

An extended material of valence electron spatial correlations (VEC) had been analyzed *(Schubert, 1964)*, when it became apparent that one correlation of valence electrons alone is not sufficient for the explanation of crystal structures of metallic phases. The outer core electrons had to be taken into consideration. This may best be seen from the crystal structure of indium (Fig. 4): The lattice matrix of In may be given in diagonal form: $a_{In} = (4.59; 4.59; 4.95)$ Å. The explicit lattice constants are needed for verification that the proposed VEC is acceptable. The VEC is $a_{In} = a'_{A1}(1, -1, 0; 1, 1, 0; 0, 0, 3/2)$ and may be decomposed into the equations $a_1 = a'_1 + a'_2$ $a_2 = - a'_1 + a'_2$ $a_3 = 3\,a'_3/2$, which may be verified by means of Fig. 4. If a correlation lattice is inserted into a crystal structure, this does not mean that there are positions of increased electron density in the cell, it only gives the commensurability which is favorable energetically. It is easily verified that the number of valence electron places per cell is $N_{P'}^{/C} = 12$ and is equal to the number of valence electrons in the cell given above $N_V^{/C} = 12$. The A1 type of the VEC had been inferred from the diamond struc-

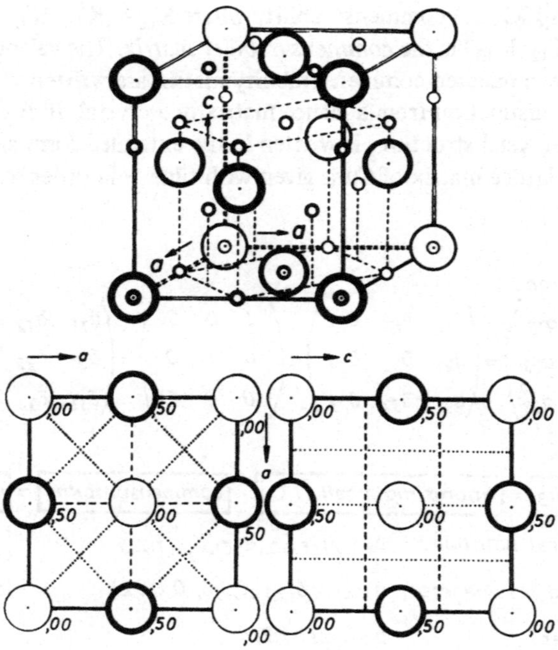

Fig. 4. Two electron spatial correlations in Indium, *broken line:* VEC; *pointed lines:* CEC.

ture of grey tin where the A1 VEC is especially natural; the comparison of both structures leads to the given assumption for the VEC. The proposed VEC gives a very good reproduction of the observed axial ratio $|a_3|/|a_1|$, when the a'_{A1} is assumed to be undistorted; but the shear variants of Cu_3Au, which will be considered below, showed a quite strong contraction of the VEC under the influence of the crystal structure. This can only be understood when we assume that there is an influence on the structure of the outer core electrons. The simplest assumption is that also the outer core electrons form a spatial core electron correlation (CEC) similar to a lattice, which is in so far independent of the VEC that the overall electron correlation may be written as a product. A remaining dependence of the two partial correlations might be sought in some commensurability between VEC and CEC. This commensurability leads to the number of electrons taking part in the CEC and to the type of correlation: in the structure of In only the 4 d^{10} electrons had to be taken into account and the type must be of the A2 lattice: $a_{In} = a''_{A2}(2, -2, 0; 2, 2, 0; 0, 0, 3)$: the CEC is distinguished from the VEC by two primes. The commensurability between VEC and CEC is here simply the factor two, which is energetically favorable, and may be called a *factorial commensurability*. An interesting feature is that this CEC gives a number of places per cell of $N^{/C}_{p''} = 48$, while only 40 4d electrons are in the cell: the CEC is only partially occupied, or stated otherwise, the number of CEC places per atom is $N^{/A}_{p''} = 12 > 10 = N^{/A}_{d}$. Because of its frequent occurrence, this binding may be named the *In binding*. Although this binding occurs also in the diamond family, it was the

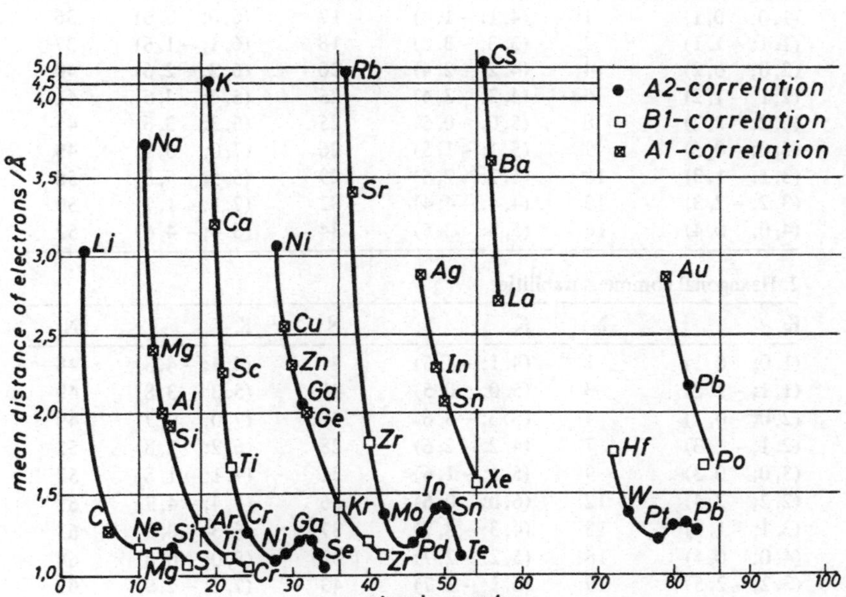

Fig. 5. Electron distances in dependence of composition.

structure of In which yielded this remarkable spatial correlation of the valence- and outer core electrons. Phases with filled outer d-shells of the components may be assumed to have nearly all an A2 CEC; they are called *brass-like phases.*

To find a spatial correlation proposal for a given phase, the number of valence electrons or of outer core electrons must fit into a correlation which is commensurable with the unit cell. A system of electron distances derived from spatial correlations of the chemical elements is often helpful Fig. 5 *(Schubert, 1964)*. With such distances several correlation types may be metrically determined and then compared with the lattice constants of the crystal and the electron number in the cell. Good commensurability to the unit cell and atomic sites is the main indication of a probable spatial correlation. For tetragonal and hexagonal cells a number of possible commensurabilities are given in Table 1. Several *empirical rules* have been found useful during the analysis of the correlations. These rules summarize observations found during the search for spatial correlations, they must be verified by the whole body of interpretations of crystal structures by the present model. In the following paragraphs many examples for the rules may be found.

Table 1. Several plane commensurabilities.

1. Tetragonal commensurabilities

K	N	K	N	K	N
(1, 0; 0, 1)	1	(4, 1; – 1, 4)	17	(6, 0; 0, 6)	36
(1, 1; – 1, 1)	2	(3, 3; – 3, 3)	18	(6, 1; – 1, 6)	37
(2, 0; 0, 2)	4	(4, 2; – 2, 4)	20	(6, 2; – 2, 6)	40
(2, 1; – 1, 2)	5	(4, 3; – 3, 4)	25	(5, 4; – 4, 5)	41
(2, 2; – 2, 2)	8	(5, 0; 0, 5)	25	(6, 3; – 3, 6)	45
(3, 0; 0, 3)	9	(5, 1; – 1, 5)	26	(7, 0; 0, 7)	49
(3, 1; – 1, 3)	10	(5, 2; – 2, 5)	29	(5, 5; – 5, 5)	50
(3, 2; – 2, 3)	13	(4, 4; – 4, 4)	32	(7, 1; – 1, 7)	50
(4, 0; 0, 4)	16	(5, 3; – 3, 5)	34	(6, 4; – 4, 6)	52

2. Hexagonal commensurabilities

K	N	K	N	K	N
(1, 0; 0, 1)	1	(4, 1; – 1, 5)	21	(4, 4; – 4, 8)	48
(1, 1; – 1, 2)	3	(5, 0; 0, 5)	25	(5, 3; – 3, 8)	49
(2, 0; 0, 2)	4	(3, 3; – 3, 6)	27	(7, 0; 0, 7)	49
(2, 1; – 1, 3)	7	(4, 2; – 2, 6)	28	(6, 2; – 2, 8)	52
(3, 0; 0, 3)	9	(5, 1; – 1, 6)	31	(7, 1; – 1, 8)	57
(2, 2; – 2, 4)	12	(6, 0; 0, 6)	36	(5, 4; – 4, 9)	61
(3, 1; – 1, 4)	13	(4, 3; – 3, 7)	37	(6, 3; – 3, 9)	63
(4, 0; 0, 4)	16	(5, 2; – 2, 7)	39	(8, 0; 0, 8)	64
(3, 2; – 2, 5)	19	(6, 1; – 1, 7)	43	(7, 2; – 2, 9)	67

K = commensurability matrix, N = number of small meshes in the greater mesh, $\sqrt{N} = |a| / |a'|$ where $|a|$ = lattice constants.

1. The valence electron concentration may be calculated from the positive valences of the component atoms, e.g. Ni^0, Cu^{1+}, Zn^{2+}, Ga^{3+} etc., the valence electron correlation is fully occupied in low valence electron concentration phases. (The possibility of partial occupation had been mentioned above.)

2. The core electron correlation is generally built of the outer d electrons of the atoms, alternatively outer sp electrons may take part. It is mostly occupied to 70 ... 90%, since the core electrons are much more localized around the atomic center.

3. The occupation of the core electron correlation is often higher in compounds of heavier elements, or stated otherwise: $N_{p''}^{\zeta A}$ decreases with increasing average atomic weight; this occurs since the d electrons of heavy atoms are pressed outward by the other core electrons.

4. Application of an external pressure causes transformations to structures with higher core electron correlation occupancy. This is found since the compressibility of the valence electron gas is greater than that of the core electron gas.

5. A high valence electron concentration favors a low CEC occupancy because of loosening of structure by the pressure of the VEC. This rule is some kind of reversal of rule 4.

6. Commensurability may be in one spatial direction non-whole numbered. Such a correlation is evidently next-neighbored to a fully commensurable correlation.

7. Factorial commensurability between valence and core electrons is energetically favorable, it often leads to an elevated melting temperature. If on the other hand we have a high melting temperature it is probable that the correlations are energetically favorable.

8. The electron distances of the compounds of a mixture form a smooth function in dependence of the mol fraction of these compounds. In a correlation of smaller coordination of the electron places the electron distance is smaller. This rule is paralleled by a rule for crystal structures.

9. The type of the correlations is mostly the A1, A2, A0 = cubic primitive AH = hexagonal primitive or C11 type (cf. abbreviations at the end of the paper), but it may be deformed by other influences. These types are all translation lattices, and the proposal of translation lattices with uniform distances is the simplest conceivable possibility for a spatial correlation.

10. The core electron distances may be strained by a high valence electron concentration i.e., by a high valence electron pressure on the core electron correlation (Fig. 6).

The above-mentioned rules should be satisfied by a correlation proposal when it is to be considered as probable. Generally the analysis leads to a first proposal for one of the phases in a mixture of elements and this proposal helps in finding proposals for neighboring phases. At this stage of the analysis the characteristic slope of the electron distance in function of the mol fraction is helpful for finding further interpretations. Often the geometrical conditions are so close to the structural findings that there can be no doubt that the analysis of the spatial correlation of the electrons is an important new means for the energetical interpretation of metallic phases.

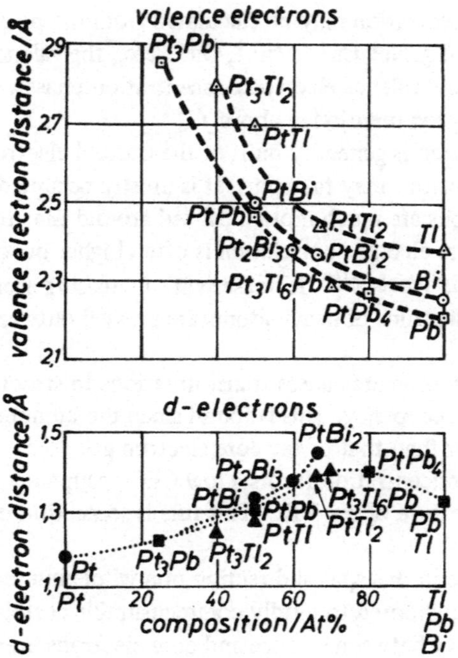

Fig. 6. Electron distances in dependence of composition (example of Pt alloys).

It seems that for a great class of metallic and nonmetallic phases the consideration of two spatial correlations (named simply a *binding*) is sufficient for the assessment of phase stability. It is therefore appropriate to speak of a *two-correlations model* which gives the first valence concept for alloys since we know chemical compounds in metallic mixtures.

It must be mentioned here that the present analysis of the extended empirical material of determined crystal structures gives moreover points of view unknown to present-day metal theory. There the valence electron gas is exclusively considered and the core electrons appear only under the concept of a core potential, i.e., spatial correlations are neither considered in the core electron population, nor is lattice-like interaction with the valence electrons; following the results of the present analysis these phenomena are quite important.

In the next paragraphs as an example for the binding model above two structural families will be considered: the Cu-family and the W-family. A *structural family* is a set of related structural types which implies related binding. The different *types* are designated by the compound first found in that type, by the type description, and by a reference. The *type description symbol* is composed of a character for the Bravais-group of the compound (cf. abbreviations at the end of the paper), of the number of atoms of the first component in the primitive cell seperated by a point from the number of atoms of the second component in the primitive cell (cf. *Schubert, 1964*). The *reference* is given for the Structure Reports e.g., Structure Report, Vol. 1, p. 13 is abbreviated to SR1.13. The reference refers to the finding of the structure not to the best parameter values; the reference only replaces the longer reference to the original literature, and does not imply that the report is entirely correct or complete. When the report of a publication is missing in SR the reference to the original literature is given in an easily understandable abbreviation. Many metallic crystal structures may be understood easily by specifying substructure (i.e., family), type description, and commensurability to the substructure; for more complicated structures the full description in the literature should be consulted (cf. e.g., *Schubert, 1964 or Pearson, 1972 or Structure Reports, 1931*. . .).

II. Cu-Family

1. Simple Substitutional Variants

In order to find a *closest packing of atoms* (of one kind), in which all atoms are translatory equivalent, we establish with the aid of the shortest atomic distance an atomic pair; because of the translatory equivalence the pair leads to a straight chain. Putting a further atom as close as possible to the chain gives an equilateral triangle and further on a lattice plane of hexagonal symmetry. Putting one more atom as close as possible to that plane gives a regular tetrahedron and further on a spatial lattice which has a cubic face-centerd symmetry. This structural type was first found in Cu and is therefore called the Cu type:

Cu(F1, SR1.13). The structure is mostly described in a cubic cell a_{Cu} which contains four atoms. A structural type has generally not only one type of binding; it may be subdivided into different subtypes, for each of which the binding is different. *Ni subtype:* $a = a'_{A2} = a''_{A2}(2, -2, 0; 2, 2, 0; 0, 0, 2.83)$ where a''_{A2} must be twinned, i.e., it must have its quasitetragonal axis statistically in all possible directions of a_{Ni}. *Cu subtype:* $a = a'_{A1} = a''_{A2}(2, -2, 0; 2, 2, 0; 0, 0, 2.83)$. *Al subtype:* The VEC may be assumed as $a = a'_{A1}(1, -1, 0; 1, 1, 0; 0, 0, \approx 1.5)$ it must also be considered as twinned; the relative stability of the alloy $Zn_{60}Al_{40}$ is to be interpreted by the fact that the VEC is no more distorted. In the structure of *In* the atomic correlation is deformed tetragonally in order to have the electron correlation given above undistorted. For the core electron correlation of Al it is useful to choose (in analogy to Mg) the hexagonal cell $a_{A1.H}$ with three atoms; if we use also a hexagonal cell $a_{A0.H}$ of a cubic primitive correlation we find a possible commensurability $a_{A1.H} = a''_{A0.H}(1, 1, 0; -1, 2, 0; 0, 0, 10.4/3)$. In the *Pb subtype* we may assume the VEC $a = a'_{A2}(2)$. In the *Ne subtype* it may be assumed $a = a'_{A0}(4)$ which would be occupied to 50%; the CEC has here no more influence. The fact that a good commensurability belongs to each subtype cannot be purely accidental.

We now consider several simple replacement variants and compare their cells with a weakly deformed cell $\tilde{a}_{Cu}$. A structure with maximum contact of different components is *Cu_3Au*(C3.1, SR1.486) $a = \tilde{a}_{Cu}$. It is found empirically that the minority component mostly has a somewhat greater atomic radius than the majority component; if it is smaller, then displacive variants occur which will be discussed below. The *Cu_3Au subtype* corresponds to the binding of Cu. The *Mg_3In subtype* is connected with the binding in Mg. The subtype of *La_3Al* and *$TiZn_3$* corresponds the Al subtype and *$NaPb_3$* to the Pb subtype of the Cu type. Shear variants and double shear variants of Cu_3Au are considered later.

We now mention several superstructures in the equiatomic region.
PtCu(R1.1, SR1.485) $a_{PtCu} = \tilde{a}_{Cu}(2)$ contains 32 atoms or 16 primitive PtCu cells. A hexagonal cell is $a_{PtCu.H} = (2.70, -1.35, 0; 0, 2.34, 0; 0, 0, 12.9)$ Å the axial ratio $|a_3|/|a_1|$ is subideal, as is to be expected from the packing of great Pt with small Cu in alternating planes of $(111)_{Cu}$. The VEC corresponds to that in Ni.

CuAu.r(T1.1, SR1.505). Mostly the doubled cell $a_{CuAu} = \tilde{a}_{Cu}$ is chosen, in which (001) planes are alternatively occupied by Cu and Au. The ordering of atoms of different size in alternating planes is energetically favorable; the distance in a plane is given by the greater atom and the distance in the normal is decreased as compared with the cubic structure; therefore the normal should be a direction of minimum elastic modulus of the VEC; it is $[001]_{Cu}$. In PtCu there is probably an ordering of the spins in the VEC so that the contraction direction becomes $[111]_{Cu}$. The CEC of CuAu probably becomes of the C11 type: for $a = a''_{C11}(2, -2, 0; 2, 2, 0; 0, 0, 3)$ the ideal axial ratio becomes $|a_3| / |a_1| = 0.87$; it seems that this ratio is preferred especially if the radius ratio of the components is approximately 1. The *MnNi* subtype corresponds to Ni; the *NiZn* subtype corresponds to Cu; the *TiAl* and *TiHg* subtype corresponds to Al, probably also the *LiBi* subtype. Superstructures of the CuAu type are:

PdCu₂Au(T1.2.1, SR19.135) $a = \tilde{a}_{Cu}$, but the C-centering of CuAu is lost.
Pt₂ZnCd(T2.1.1, JLCM 20.1970.266) $a = \tilde{a}_{Cu}(1; 1; 2)$, Zn and Cd planes alternate with Pt-planes.
Mn₁₁Pd₂₁(Phil.Mag.16.1967.1063).
CuAu.h1(Q10.10, SR23.136) $a = a_{CuAu.r}(1; 10; 1)$, shear variant of CuAu; shear variants of Cu_3Au will be described below.

Further superstructures of the Cu type.
Pt₇Mg(F7.1, SR27.322) $a = \tilde{a}_{Cu}(2) = a'_{A0}(2)$. Curiously Ca_7Ge(SR 28.14)is isotypic.
TiPt₈(U1.8, SR30.86) $a = \tilde{a}_{Cu}(1.5, -1.5, 0; 1.5, 1.5, 0; 0, 0, 1) = (8.312; 8.312; 3.897)$ Å. The binding may be assumed as $a = a'_{A2}(2; 2; 1) = a''_{A2}(6; 6; 2.8)$, a seldom case of 3-factorial binding, (Fig. 7). The fact that this structure does not occur in the

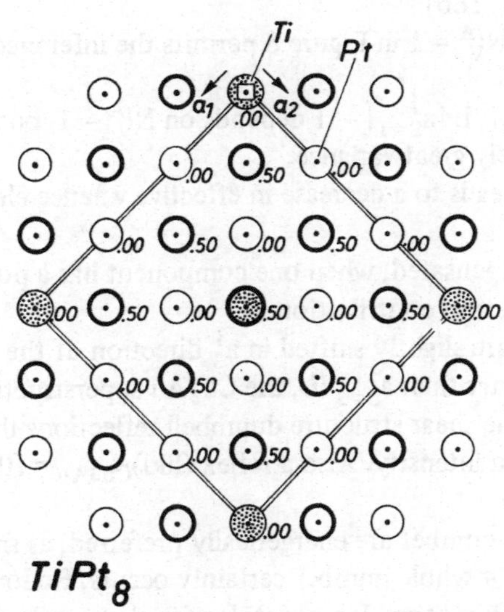

Fig. 7. Structure of $TiPt_8$.

mixture Pt-Ge is compatible with the observation that the ionization energies are lower for Ti. $NbNi_8$ is isotypic to $TiPt_8$ when a critical number of vacancies has been produced (Trans. MSAIME 245 (1961) 345).
(The displacive superstructures of the Cu type are considered later.)

2. *Variants with Superstructure Shearing*

If $a_{Cu3Au}: = a_1^i, a_2^i, a_3^i$ are the edge vectors relative to the coordinate system i of the elementary cell of Cu_3Au, we find after alloying several percent of Zn in $Cu_{3-m}AuZn_m$ (where m is a marginal value) a new cell $a_S = a_{Cu3Au}(1; 1; L)$ of a *shear superstructure*, where $K_{33} = L = 2 \ldots 10$ is the *shear length*. If $f_{Au} * \delta(x, y, z)$ means that an Au atom is located at the fractional coordinate x, y, z and if L/2 is whole numbered, we may describe the Au positions in a_S^i by $f_{Au} *(\delta(0, 0, 0/L) + \delta(0, 0, 1/L) + \ldots + \delta(1/2, 1/2, L/2\,L) + \delta(1/2, 1/2, (L/2 + 1)/L) + \ldots)$. $a_{S=3}^i$ is the *shear normal*, $(a_{S=1}^i + a_{S=2}^i)/2$ is the *shear vector*; since it is normal to $a_{S=3}^i$ we speak of a *normal shear structure*. The plane orthogonal to $a_{S=3}^i$ between the Au sites $\delta(0, 0, (L/2 - 1)/L)$ and $\delta(1/2, 1/2, L/2\,L)$ is called the *shear plane*. The reciprocal lattice of the shear structures above may be constructed by multiplying the diffraction amplitudes of a few Cu_3Au cells with the diffraction amplitude of a tetragonal body-centered a_S^i lattice of δ functions. The following properties of shear structures have been observed:

1. Shear structures occur in many brass-like alloy systems near the valence electron concentration $N_V^{/A} = 1$. The *shear density* $1/L =: D_S$ depends on $N_V^{/A}$ approximately by $D_S = N_V^{/A} - 1$, Fig. 8.

2. This equation gives a better fit to the observation if an *effective concentration* $N_V^{eff} > N_V^{/A}$ is used (JLCM 37, 1974, 166).

3. The *cut-off-interval* around $N_V^{/A} = 1$ in Figure 8 permits the inference of the range of binding forces as 50 Å.

4. The *axial ration strain* $|a_{S=3}^i| / L\, |a_{S=1}^i| - 1$ depends on $N_V^{/A} - 1$, both having the same sign; but there is a relatively great variance.

5. An increase in temperature leads to a decrease in effective valence electron concentration.

6. This effect may be over-compensated, when one component has a positive temperature dependence of the electron contribution.

7. Atoms near the shear plane are slightly shifted in a_3^i direction. If the shears make the Cu_3Au type subcells shorter than $a_{S=3}^i/L$, the Cu_3Au superstructure reflex goes to greater θ so that in the shear structure dumbbell reflections the component with greater θ is increased in intensity. Also a reflex $(hkl)_{Cu3Au} = (0, 0, 2\,D_S)$ appears.

8. Structures with L/2 = whole number are energetically preferred, as small plateaus of $D_S(N_V^{/A})$ show; but L/2 ≠ whole number certainly occurs. Following *Fujiwara* (1957) a dumbbell series (h = const, k = const, l) with sharp reflexions is caused for L/2 ≠ whole number by a structure amplitude $F_l = \Sigma_n^{0 \ldots N-1} f_n \exp 2\pi i n l$

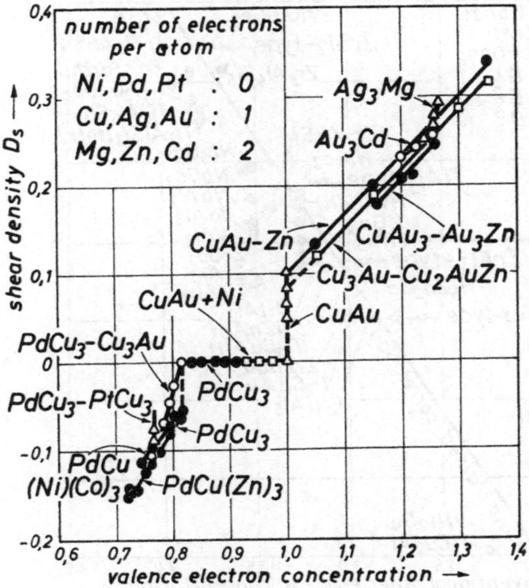

Fig. 8. Shear variants of the Cu_3Au type.

with f_n = (1 for Lm $\leqslant$ n < L(m + 1/2), – 1 for L(m + 1/2) $\leqslant$ n < L(m + 1)) where m = whole number and N = great number; the unsheared substructure cell sequences of approximate length L/2 represent lengths of minimal iteration. If L is a fraction of low whole numbers, a small superstructure cell is found giving additional reflecions.

9. Besides the shear variants of Cu_3Au in the neighborhood of $N_V^{/A} = 1$ another set is found near $N_V^{/A} = 3$ represented by $TiAl_3$(U 1.3, SR 16.124) and $ZrAl_3$(U 2.6, SR 7.14). Ternary alloys show $TiAl_3$ representatives to be richer in valence electrons than $ZrAl_3$ representatives. The axial ratio is a function of $N_V^{/A}$ (Fig. 9).

10. Shear variants of the Cu_3Au structure with two systems of shear planes will be treated later.

Energetic explanations of these properties have been given by the band model (*Sato* and *Toth,* 1965) and by the spatial correlation model (*Schubert,* 1964); the latter model explains more facts: In Cu_3Au an A1-VEC $a_{Cu3Au} = a'_{A1}(1;1;1)$ may be assumed. If $N_V^{/A}$ is increased by substituting some Zn for Cu, the number of electron places per cell $N_{P'}^{/C}$ must also be increased. Since the above commensurability between VEC and atomic correlation (AC) is energetically favorable, it is profitable to loosen commensurability only in one direction by inserting further electron planes perpendicularly to that direction. As by this insertion the crystal will be deformed, the abovementioned direction must be of minimum elastic modulus: it is a_3^i. By the incommensurability between AC and VEC a wave-like distribution of electric dipole vectors arises at the atoms, which will be different at the minority component of the struc-

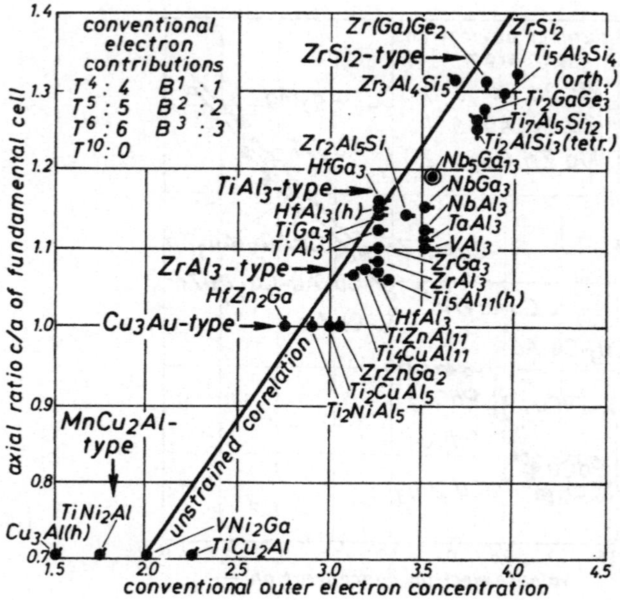

Fig. 9. Axial ratio of structures homotypic to $TiAl_3$.

ture from the dipole vector at the majority component (Fig. 10). If the VEC is given by $a_S = a_{A1}(1; 1; L + 1)$ just one wavelength of the dipole array in the a_3^i direction is completed; the shearing helps minimize the energy of the array. It follows $4\ LN_V^{/A} - 4\ L = 4$, so that relation (1) is fulfilled. Since several electron places may be occupied by atom cores, the effective valence electron concentration, which has

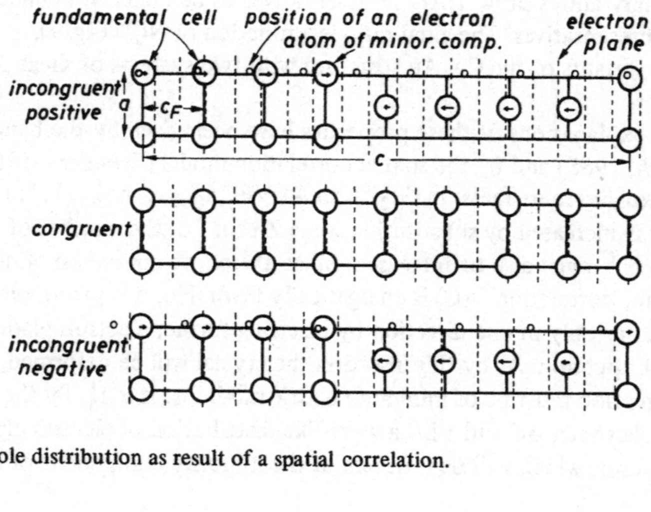

Fig. 10. Dipole distribution as result of a spatial correlation.

to do with a fully occupied VEC, is greater than the nominal one (2). The spatial range of the binding forces (3) must be comparable to the periodicity radius of the VEC. The sign of $|a_3|/L|a_1| - 1$ comes from the fact that electron planes orthogonal to a_3 are added or subtracted according to the sign of $N_V^{/A} - 1$ (4); it is simplifying to count D_S negative in the case $N_V^{/A} - 1 < 0$. An increase of temperature leads to a population of unoccupied electron places, therefore N_V^{eff} becomes more similar to $N_V^{/A}$. An example for temperature-dependent VE-contribution (6) is given by Au. Following present knowledge the displacements (7) may be described by the rule that the atom richer in valence electrons moves to the shear plane according to the $|a_3|/L|a_1|$ change; this corresponds to the expectation derived from the strain in the electron lattice. The rule of *Fujiwara* (8) corresponds to the energetic influence of the dipole distribution. The $TiAl_3$ subtype of the shear variants represents the next electron rich commensurability in the tetragonal basal plane: $a_{TiAl3} = a'_{A1}(1,-1,0; 1,1,0; 0,0,3.5)$ and $a_{ZrAl3} = a'_{A1}(1,-1,0; 1,1,0; 0,0,6.5)$; here each electron plane perpendicular to a_3 contributes to the dipole array; as here more electrons are involved in the VEC than in $Cu_{3-m}AuZn_m$ the VEC is less strained than in $Cu_{3-m}AuZn_m$; the number $N_{P'}^{/C'}$ is greater for $TiAl_3$ than for $ZrAl_3$ if we choose C′ as the substructure cell, conforming to the rule, that in heavier compounds the occupation of the electron correlation is higher; this comes from the fact that in heavy compounds the outer electrons are more pressed away from the atom by the inner electrons.

A special case of shear variants of Cu_3Au is given by the compound Pd_3Tl consisting of two phases: $Pt_3Tl \cdot r$ ($ZrAl_3$ type, SR34.141), a = (4.117; 4.117; 15.274) Å, $|a_3|/4|a_1| = 0.93$ and $Pd_3Tl \cdot h$ ($TiAl_3$ type, SR33.125), a = (4.146; 4.146; 7.497) Å, $|a_3|/2|a_1| = 0.90$. The VEC for $Pd_3Tl \cdot r$ $a = a_{A1}(1; 1; 3)$ favors the $ZrAl_3$ type. If $a_{Pd3Tl \cdot h} = a''_{A2}(2,-2,0; 2,2,0; 0,0,5)$ and $a_{Pd3Tl \cdot r} = a''_{A2}(2,-2,0; 2,2,0; 0,0,10.5)$ it is to be understood that in the h-phase the $TiAl_3$ type is preferred, under the condition that few valence electrons enter the 5f niveau so that the VEC becomes $a_{Pd3Tl \cdot h} = a'_{A0}(1,-1,0; 1,1,0; 0,0,2.5)$.

3. *Anormal and Multiple Superstructure Shears*

An interesting set of structures is given by a representative of the shear variants of Cu_3Au, the *VNi_3*($TiAl_3$ type, SR 16.124) (Fig. 11) a = (3.54; 3.54; 7.21) Å = $\tilde{a}_{Cu}(1; 1; 2) = a'_{A2}(1,-1,0; 1,1,0; 0,0,2.9 \approx 3) = a''_{A2}(2,-2,0; 2,2,0; 0,0,6)$. The strained axial ratio is explained by the binding. In order to conserve this favorable factorial binding at valence electron concentrations different from $N_V^{/A} = 5/4$ in VNi_3, nature uses the mechanism of superstructure shears with shear vectors not normal to the shear normal *(anormal shears)*, the shear plane being parallel to a plane of maximum V-atom density not interfering with the primary shear system:

$MoNi_4$(U 1.4, SR 9.110) a = (5.72; 5.72; 3.56) Å = $\tilde{a}_{Cu}(1.5,-0.5,0; 0.5,1.5,0; 0,0,1)$. This cell may be obtained from VNi_3 by shears $(-101)[1/2,0,1/4]_{VNi3}$.

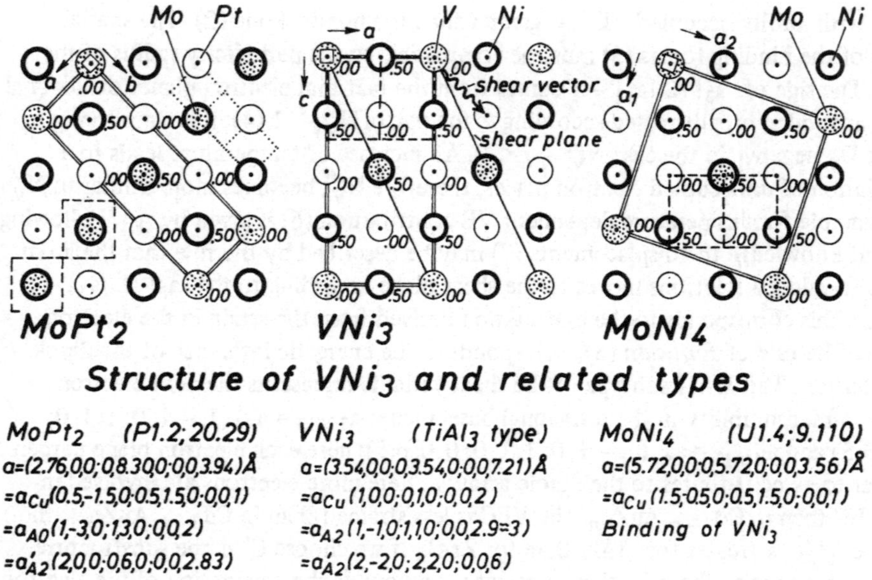

Fig. 11. Structure of VNi_3 and related types.

The binding remains as in VNi_3 and must be assumed as twinned; the short substructure axial ratio is explained by that binding. Besides the shear straining the donator position also a compressive shear (– 101) [– 1/2, 0, – 1/4] has been observed in the following phase.

$MoPt_2$(P1.2, SR 20.29) $a = (2.76; 8.30; 3.94)Å = \tilde{a}_{Cu}(0.5, 1.5, 0; -0.5, 1.5, 0; 0, 0, 1)$. This binding can no longer be of the $a'_{A2} = a''_{A2}(2)$ type, since the $N_V^{/A}$ is increased by the shear; one finds $a = a'_{A0}(1, 3, 0; -1, 3, 0; 0, 0, 2) = a''_{A2}(2; 6; 2.83)$. Since the $MoPt_2$ type occurs also in the mixture V–Ni, and since it is in equilibrium with a Cr_3Si related phase, it may be anticipated that the A0 correlation plays an important role in the Cr_3Si family.

Mn_2Au_5(N 2.5, SR 29.60) (Fig. 12) $a = \tilde{a}_{Cu}(-1, 0, 1.5; 0, 1, 0; 2, 0, 0.5)$. This structure has the same shear planes as $MoNi_4$ and $MoPt_2$, however the shear density is smaller. It appears that in this structure three correlations are of influence (*Schubert*, 1976) at any case the correlation of the Mn-electrons explains the Mn dumbbells. Another member of this family, the $TiPt_8$, has been mentioned among the simple superstructures of the Cu type.

While the foregoing phases have an oblique second system of shear, structures with two orthogonal shear systems have also been observed:

Au_4Zn(O 30.90, SR 21.110) $a = \tilde{a}_{Cu}(1; 6_u; 5_n)$. A normal shear system (n) has the normal a_3 and the density 1/5, an anormal shear system (u) has the normal a_2 and the density 1/6. The anormal shear makes possible the off stoichiometry composition of

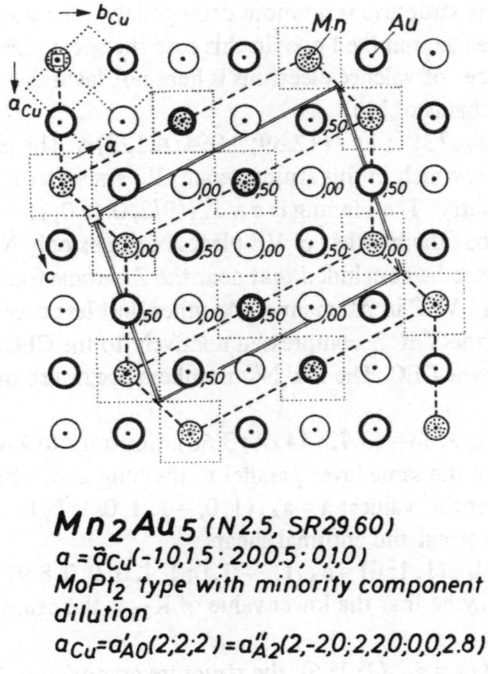

Fig. 12. Structure of Mn_2Au_5.

the phase. The normal shear density conforms to the $D_S(N_V^{(A)})$ relation of brass like shear structures and is to be explained in the given manner; the anormal shear system contributes to lowering the energy of the dipole array in the case of off-stoichiometry. Curiously enough a structure with two normal shear system has been observed: $Cu_{.50}Au_{.20}Zn_{.30}$(O 16.48, SR 21.110) a = $\tilde{a}_{Cu}(1; 4_n; 4_n)$. Double shear structures have also been found with negative shear densites:

$Pd_{1+}Cu_{3-}$.*r*(O 48.144, SR 19.146) a = $\tilde{a}_{Cu}(8_u; 1; 6_n)$. A normal shear system has the shear normal a_3 and an anormal system the shear normal a_1. In this case also the modulus of the anormal density is smaller than the modulus of the normal density.

$Cu_{\sim .40}Au_{\sim .60}$(M 128.192, JAC 8.1975.598) a = $\tilde{a}_{Cu}(10_n, 1, 0; 0, 8_u, 0; 0, 0, 1)$, it appears that the anormal density is higher than the normal one. Also double shears without off-stoichiometry have been found:

$MnAu_3$(O 8.24, SR 24.148) a = $\tilde{a}_{Cu}(2_n; 4_n; 1)$; the structure contains two normal shear systems. In this and in related phases magnetic superstructures have been found (e.g., SR 27.220).

Shear superstructures of the anormal or multiple kind in the Al-subfamily of the Cu structure family are numerous. If the composition is $N_2' < 0.75$, a shear variant of the Cu_3Au type remains possible in the Al-subfamily when the shear vector which shifts a mixed atom layer in the structure is no more orthogonal to the shear normal so that the mixed layer replaces an unmixed one; in this case the shear plane is $(001)_{TiAl3}$; probably the number of valence electrons is here not favorable to the $MoPt_2$ type, with its straight chains of Mo.

$ZrGa_2$(Q 2.4, SR 27.205) $a = \tilde{a}_{Cu}(3; 1; 1) = (12.89; 3.994; 4.123)$Å. The anormal shear forms zigzag chains of Zr, which in this structure are all parallel to a_2; the chains cause the orthorhombic symmetry. The binding is $a = a'_{A1}(9/2, 0, 0; 0, 1, -1; 0, 1, 1) = a''_{A2}(9, 0, 0; 0, 2, -2; 0, 2, 2)$, but the number of VE places becomes only $N_{P'}^{/C} = 36$ while $N_V^{/C} = 40$. It must therefore be concluded that near the Zr atoms four electrons per cell are inserted into the A1 VEC in A0 manner. Another, but less probable, possibility is the assumption that the VEC is compressed relatively to the CEC; or that several valence electrons enter the CEC. The low $N_{P'}^{/C}$ numbers occur also in the homologous structures.

$ZrSi_2$(Q 2.4, SR 5.5) $a = \tilde{a}_{Cu}(1; 3; 1) = (3.72; 14.68; 3.68)$. Contrary to $ZrGa_2$ the Zr chains are parallel but not in the same layer parallel to the long axis, which may be understood from the different K′ values: $a = a'_{A1}(1, 0, -1; 1, 0, 1; 0, 11, 0)$. This structure shows a mixture of normal and anormal shears.

$HfGa_2$(U 4.8, SR 27.207) $a = \tilde{a}_{Cu}(1; 1; 6) = a'_{A1}(1, -1, 0; 1, 1, 0; 0, 0, 8.9) = a''_{A2}(2, -2, 0; 2, 2, 0; 0, 0, 17.7)$. It may be that the lower value of K'_{33} is the cause of the Hf chains alternating in direction.

$Zr_{1-}AlSi_{1+}$(U 4.8, SR 30.108) $a = \tilde{a}_{Cu}(1; 1; 6)$, the structure compares to $ZrSi_2$ like $HfGa_2$ to $ZrGa_2$. The binding may be assumed as $a = a'_{A1}(1, -1, 0; 1, 1, 0; 0, 0, 22.4)$ similar to $ZrSi_2$.

Nb_5Ga_{13}(Q 5.13, SR 30.139) $a = \tilde{a}_{Cu}(1; 1; 9)$. The structure is composed of elements of the $TiAl_3$ type and $ZrSi_2$ type. The VEC is $a = a_{A1}(1, -1, 0; 1, 1, 0; 0, 0, 30.6)$ giving $N_{P'}^{/C} = 124$ while $N_V^{/C} = 128$. Here also few valence electrons are in a special condition.

Together with these structures the TlJ type of ZrAl(SR 27.37) should be seen, or the MoB type of ZrGa.r(SR 27.205) $a = \tilde{a}_{Cu}(1; 1; 4) = (3.865; 3.865; 20.56)$Å $= a'_{A1}(1, -1, 0; 1, 1, 0; 0, 0, 7.5)$. The great strain in the a_3 direction makes the layers possible with stacking by the support number 1.

Double shear structures with different direction of the shear planes of the A1-family have also been found:

Ti_2Ga_3(T 4.6, SR 27.206) $a = \tilde{a}_{Cu}(1.5, 0.5, 0; -0.5, 1.5, 0; 0, 0, 1)$. Two anormal shear systems produce the 2/3 stoichiometry. The shear system of $MoNi_4$ has also been observed here:

Mo_3Al_8(N 3.8, SR 27.377) $a = \tilde{a}_{Cu}(2, 0, 0.5; -1, 0, 2.5; 0, 1, 0)$. Since the Mo-content is increased relative to the 1/3 stoichiometry, it is to be assumed that the A1 correlation of A1 is filled in the neighborhood of the Mo to an A0 correlation: $a = a'_{A1}(1, 1, -2.5; -1, 1, 2.5; 3.5, 0, 0.87)$ which takes $N_{P'}^{/C} = 70$ places; since $N_V^{/C} = 84$ it must be assumed that per Mo-dumbbell 7 valence electrons fill the A1 correlation in A0 manner; it is easily seen that this gives a good neighborhood for the Mo dumbbells.

4. *Displacive Variants*

If many displacements (other than shears) of atoms away from ideal positions are observed in an elementary cell, which may not be described by a homogenous strain, we speak of a *displacive structure*. If in a Cu_3Au-related phase the minority atom has a greater radius than the majority atom, a Cu_3Au structure is found; but if the minority atom is smaller, then in the polyhedron of next neighbors an instability arises which may lead to an inhomogenous system of displacements *("displacion")*. It is to be assumed that the displacements are caused by the CEC, while shearing was caused by VEC. Since the CEC is to some extent independent of the VEC, in a displacive structure shearing is also possible. The energetic weight of these kinds of ordering may be assessed from the Au–Zn phase diagram: in the compound Au_3Zn down to 415 °C a Cu type is stable, between 415 and 260 °C a $ZrAl_3$ type of shear structure, and from 260 °C down to room temperature the structure becomes displacive; it may be concluded that the energy decrease during the shear transformation is greater but comparable to the energy decrease during the displacive transformation. We consider first representatives of displacive structures based on the composition 3/1. *Ir_3Si*(U 6.2, SR 24.113) (Fig. 13), $a = \tilde{a}_{Cu}(1, 1, 0; -1, 1, 0; 0, 0, 2) = (5.22; 5.22; 7.95)$ Å. The position of the Si atoms corresponds to the Au-position in Cu_3Au. The small Si-core permits four of its Ir-neighbors to come nearer to it than the rest of the nearest neighbors. The binding is: $a_{Ir3Si} = a'_{A1}(1, 1, 0; -1, 1, 0; 0, 0, 2) = a''_{A2}(4; 4; 6)$. *$Pt_3Ga$. r*(T 12.4, *Chattopadhyay* and *Schubert*, 1975). The structure is quite similar to Ir_3Si, but there is no body-centering. The position of the Ga shows no shearing

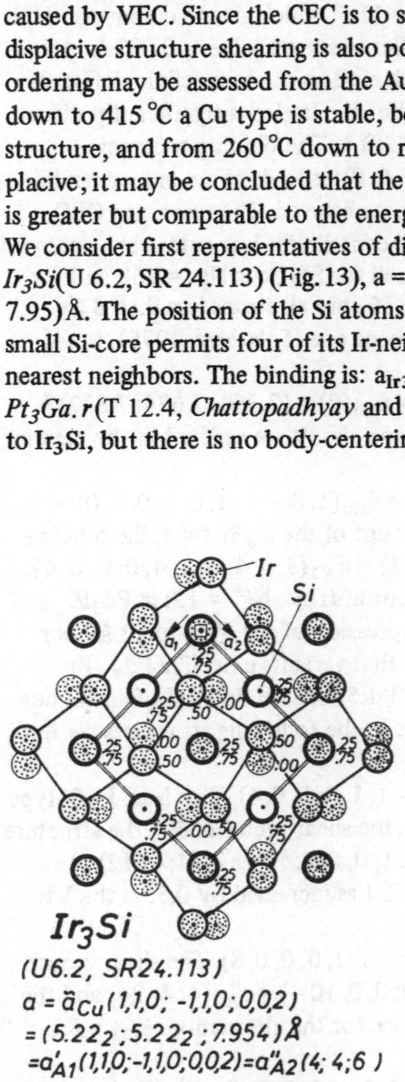

Fig. 13. Structure of Ir_3Si.

(while Pd_3Tl shows shear structures). It is therefore to be assumed that the VEC is similar to that in Ir_3Si, and that several VEP are occupied by Pt-cores. For the large Tl-atoms such a compression of the VEC is not possible, therefore a shear structure is stable.

Pt_3Ge.r(N 6.2, SR 24.116) $a = (7.922, 0, 0; 0, 7.768, 0; -0.087, 0, 7.768)$ Å $= \tilde{a}_{Cu}(2; 2; 2)$ the lattice is here not given for the N 6.2-cell but for a quasicubic on all faces centered cell with 32 atoms. The position of the Ge is like the Au position in Cu_3Au, which is compatible with the VEC $a = a'_{A1}(2)$. The monoclinic distortion is explained by the CEC: $\hat{a} = a''_{A2}(0, 4, -4; 0, 4, 4; 5.75, 0, 0)$ where $\hat{a}$ ia thought to be an orthogonalized cell of Pt_3Ge.r. It follows that $2\,K_{31} = 11.5$, which gives a commensurability gain when $a_{31} = -0.49$ Å; the essentially smaller observed value -0.087 Å indicates that the CEC is deformed by the VEC which hinders the monoclinic deformation. The structure of Pt_3Si is isotypic to Pt_3Ge but the lattice is significantly different $a = (7.697, 0, 0; 0, 7.758, 0; -0.254, 0, 7.758)$ Å. The binding becomes $\hat{a} = a'_{A1}(2; 2; 2) = a''_{A2}(0, 4, -4; 0, 4, 4; 5.6 \approx 5.5, 0, 0)$; it leads to a commensurability gain for $a_{31} = -0.97$ Å while $a_{31} = -0.254$ Å is observed. The surprising CEC-element $K''_{31} = 5.5$ is explained by the fact that Si contributes less outer core electrons than Ge; the number of core electron places per cell for Pt_3Ge is $N^{/C}_{p''} = 368$ and for Pt_3Si, $N^{/C}_{p''} = 352$; the 8 Si-atoms contribute only 16 core electrons less than 8 Ge-atoms. An explanation of this, other than the present one (*Schubert*, 1975), is not known.

$NiPt_2Ge$(S 2.4.2, *Alasafi* and *Schubert*, 1977) $a = \tilde{a}_{Cu}(2) = (7.364; 7.674; 7.822)$ Å. The structure is similar to Ir_3Si, but the Ni + Ge and the Pt are ordered in the manner of a CuAu structure.

Pd_3Si(O 12.4, Fe_3C type, SR 24.205) (Fig. 14) $a = \tilde{a}_{Cu}(1, 0, -1; 1, 0, 1; 0, 2, 0) = (5.735; 7.555; 5.260)$ Å. This is a deformation variant of the Ir_3Si type; the binding may be assumed as $a = a'_{A1}(1, 0, -1; 0, 2, 0; 1, 0, 1) = a_{A2}(3, -4, 0; 3, 4, 0; 0, 0, 4)$. The number of C places per atom is $N^{/A}_{p''} = 11.5$ but in Ir_3Si, $N^{/A}_{p''} = 12$; in Pd_3P, which is isotypic to Pd_3Si, we may assume a compression of the VEC to an A2 correlation. The Fe_3C type is obviously so favorable that variants exist like $Pd_{4.8}P$:

$Pd_{4.8}P$(M 18.4, SR 31.54) $a = \tilde{a}_{Cu}(1, 0, -1.5; 1, 0, 1.5; 0, 2, 0)$. Instead of an A2 correlation in Pd_3P we may assume here an A1 VEC. In the following structures we find an interference of shearing and displacing.

Au_3Zn_{1+}.r(Q 12.4, SR 22.107) $a = \tilde{a}_{Cu}(1, 1, 0; -1, 1, 0; 0, 0, 4)$. This is an Ir_3Si type of structure with shears of the shear length $L = 4$; the shear vector makes the structure orthorhombic. The binding is $a = a'_{A1}(1, 1, 0; -1, 1, 0; 0, 0, 5) = a''_{A2}(4; 4; 12)$. As compared with Pt_3Ge the matrix element $K''_{33} = 12$ has increased by 0.5, as the VE exert a greater pressure on the cell.

Au_3Zn_{1-}.r(U 48.16, SR 23.155) $a = \tilde{a}_{Cu}(1, 1, 0; -1, 1, 0; 0, 0, 8)$. The shear vector alternates. The binding is $a = a'_{A1}(1, 1, 0; -1, 1, 0; 0, 0, 10-) = a''_{A2}(4; 4; 24)$ and the small defect on K'_{33} must be considered as the cause for the alternation. $K'_{33} = K_{33} + 2$ not $K_{33} + 1$ because of the displacement system.

Au_9Zn_2Ga(U 18.4.2, SR 22.108) $a = \tilde{a}_{Cu}(1, -1, 0; 1, 1, 0; 0, 0, 6) = a'_{A1}(1, -1, 0; 1, 1, 0; 0, 0, 8) = a''_{A2}(4; 4; 18)$.

In the case of the composition 2/1 the Cu_3Au ordering of the minority component is no longer possible. The following structures have been observed.
Ni_2Si(O 8.4, SR 16.123) $a = \tilde{a}_{Cu}(1, -1.5, 0; 1, 1.5, 0; 0, 0, 1) = (4.99; 7.06; 3.72)$ Å. The position of the minority component may be derived from a Cu_3Au-position by anormal shearing. The binding is $a = a'_{A2}(2; 3; 1.5) = a''_{A2}(4; 6; 3)$. The question arises why in Ni_2Si (or Pd_2Ga(Ni_2Si type)) the commensurability $K''_{33} = 3$ is accepted and in Cu not accepted. Since all valence electrons in Ni_2Si belong to one and the same component, the VEC is more easily deformed than in Cu where each valence electron belongs to another atom, therefore $K''_{33} = 3$ is more easily accepted; also the fact that Si-atoms are neighbored gives support to the distortion. In the mixture Ni–Ga no phase of the Ni_2Si type becomes stable, since the core of the minority atom is not small enough.
Rh_5Ge_3(O 10.6, SR 19.177) $a = \tilde{a}_{Cu}(1, -2, 0; 1, 2, 0; 0, 0, 1) = (5.42; 10.32; 3.96)$ Å. Here also, the Ge position may be derived by anormal shears from a Cu_3Au position. The binding is $a = a'_{A2}(2; 4; 1.5) = a''_{A2}(4; 8; 3)$; it is factorial and therefore favorable; shear variants are also known to this structure:

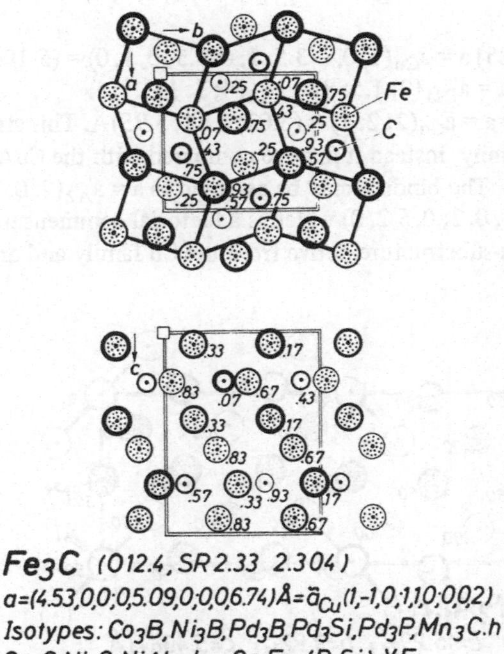

Fig. 14. Structure of Fe_3C (Isotypes Co_3B, Ni_3B, Pd_3B, Pd_3Si, Pd_3P, Mn_3C.h, Co_3C, Ni_3C, $NiAl_3$, La_3Co, $Fe_3(B, Si)$, YF_3).

Au_5Zn_3(Q 48.16, SR 22.107) $a = \tilde{a}_{Cu}(2, -1, 0; 2, 1, 0; 0, 0, 8) = a'_{A1}(2, -1, 0; 2, 1, 0; 0, 0, 11) = a''_{A2}(8; 4; 24)$. Although the A1 correlation is strongly compressed a factorial commensurability has not yet been reached. This has been found in a ternary alloy phase:
Au_5Zn_2Ga(P 10.4.2, SR 22.108) $a = \tilde{a}_{Cu}(2, -1, 0; 2, 1, 0; 0, 0, 2) = a'_{A1}(2, -1, 0; 2, 1, 0; 0, 0, 3) = a'_{A2}(4; 2; 3) = a''_{A2}(8; 4; 6)$; In this structure factorial commensurability of VEC and CEC is attained, but from the existence of Au_5Zn_3 it is to concluded that the neigborhood to factorial commensurability is already energetically favorable. Ti_3Pt_5 is isotypic to this structure and $N_V^{/C}$ are the same.
$Pt_2Al.r$(O 16.8, *Chattopadhyay* and *Schubert*, 1976; Fig. 15) $a = \tilde{a}_{Cu}(1, -3, 0; 1, 3, 0; 0, 0, 1)$. $Pt_2Ga \cdot r$ is isotypic. The occurrence of this structure is in so far interesting as the phase $Pt_2Al \cdot h$ is of the Ni_2Si type: $a_{Pt2Al \cdot h} = (5.401; 7.898; 4.055)$Å $= a'_{A1}(1, -1.5, 0; 1, 1.5, 0; 0, 0, 1) = a''_{A2}(4; 6; 3)$ or in the case that Pt contributes valence electrons $= a'_{A2}(2; 3; 1.5)$, and the r-structure where the VE contribution of the Pd will tend to zero gives the binding $a_{Pt2Al \cdot r} = (5.439; 16.297; 3.921)$Å $= a'_{A2}(2; 6; 1.5) = a''_{A2}(4; 12; 2.88)$. The VEC is not fully occupied, therefore the distribution of the A1 becomes non-uniform and K''_{33} decreases to 2.88.
Pd_5Ga_2(O 20.8, *Khalaff* and *Schubert*, 1974) $a = \tilde{a}_{Cu}(1, -3.5, 0; 1, 3.5, 0; 0, 0, 1) = (5.485; 18.369; 4.083)$Å. The binding is $a = a'_{A1}(1, -3, 0; 1, 3, 0; 0, 0, 1) = a''_{A2}(4; 13.5 \approx 14; 3)$.
Ru_4Si_3(O.16.12, SR 34.125) $a = \tilde{a}_{Cu}(1, 0, -3.5; 1, 0, 3.5; 0, 1, 0) = (5.194; 4.022; 17.136)$Å. The binding is $a = a'_{C11}(2; 1.5; 8) = a''_{A2}(4; 3; 13)$.
Pt_5Ga_3(Q 5.3, SR 24.115) $a = \tilde{a}_{Cu}(2; 2; 1) = (8.03; 7.44; 3.95)$Å. This structure does not belong to the Ni_2Si-family, instead it may be compared with the CuAu type, because of the short b-axis. The binding may be assumed as $a = a'_{A2}(2, 0, -1; 2, 0, 1; 0, 2.6, 0) = a''_{A2}(4, 0, -2; 4, 0, 2; 0, 5.2, 0)$ which is a factorial commensurability. Stacking variants of the Cu-substructure derive from the Cu family and are not considered here.

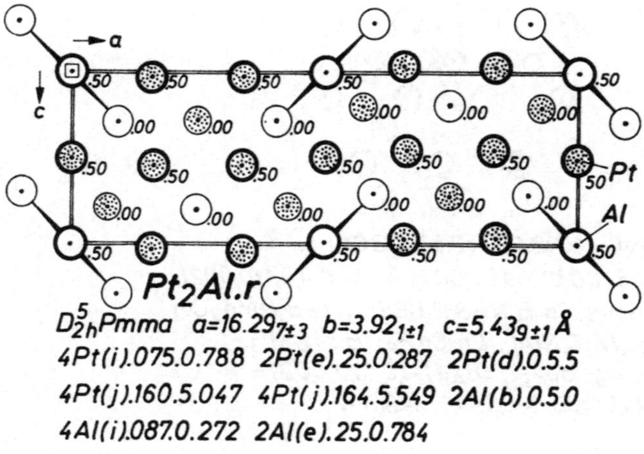

Fig. 15. Structure of $Pt_2Al.r$.

III. W-Family

1. *Cubic Substitutional Variants*

The structural type of *W*(B1, SR 1.15) is less closely packed than Cu; nevertheless Cu has only 12 neighbors having a common surface with the atomic polyhedron, while W has 14. It may be defined by a structure for a compound $A_1A'_1$, where there is maximum coordination AA′ and translatoric symmetry between all A and between all A′. Following subtypes are known. *Na subtype:* the binding $a = a'_{A2}$ is favorable as one electron avoids other electrons with parallel spins because of antisymmetry. *Be.h subtype* has the binding $a = a'_{A1}$. *V subtype:* in Ti·h we have $a = a'_{A0}(2)$ and this correlation is filled in the V- and Cr-phase in A2 manner. *Mn and Fe subtype* has perhaps the binding $a = a'_{A2}(2)$. *Tl.h subtype* and *Actinide subtype* have a binding not yet known. The *CuZn.h subtype* is mostly found at $N_V^{/A} = 1.5$ (*rule of Hume-Rothery* for beta brass). At this $N_V^{/A}$ a binding $a'_{A2} = a''_{A2}(2)$ is possible in brass-like alloys. The commensurability of this binding to the structure becomes apparent in different low-temperature structures, for instance AuCd, which will be discussed below. Several of the phases of the CuZn subtype decompose eutectoidally at fairly high temperatures. This has been explained by the low rigidity of the (110) $[1\bar{1}0]$ shear, which causes low oscillation frequencies, therefore closely neighbored phonon energies and a high entropy, so that the free energy will be lowered especially at high temperatures (Zener).

We shall now consider several replacement variants of the W-structure.
$Fe_{13}Ge_3$(C 13.3, Ann. Chim. 8.1963.85) $a = a_W(2)$. One Ge position of a "Fe_3Ge" of the Fe_3Si type is occupied by Fe.
Fe_3Si(F 3.1, SR 1.488 1.588) $a = a_W(2)$. The Si are in an F1 array in the cell. With the electron count $Fe_3^0Si_1^4$ the binding $a = a'_{A2}(2)$ follows. The *Cu_3Al subtype* belongs to the binding in CuZn; for the *Li_3Bi subtype* the binding $a = a'_{A1}$ (2) may be assumed.
$MnCu_2Al$(F 1.2.1, SR 1.488) $a = a_W(2)$, binding probably $a = a'_{A1}$ (2).
CsCl(C 1.1, SR 1.74); *VMn subtype* corresponds to V subtype; *TiFe subtype* $a = a'_{A1} = a''_{A2}(2)$; *PdCu subtype* $a = a'_{A1}(0.5, -0.5, 0; 0.5, 0.5, 0; 0, 0, 1.4/2)$. This binding explains the off-stoichiometry $Pd_{40}Cu_{60}$ of that phase; *TiZn subtype* $a = a'_{A2}(1, -1, 0; 1, 1, 0; 0, 0, 1.4)$, *CuZn.r subtype* the binding could be $a(-1, 1, 1; 0, -1, 1; 1, 0, 1) = a_{CuZn.H} = a''_{C\,11.H}(3, 1, 0; -1, 4, 0; 0, 0, 2.5)$ with $N_{P''}^{/C} = 21.7$, and $a_{CuZn.H} = a'_H(2; 2; 2.5)$, of Ni_2Al_3. In the *MgHg subtype* the binding $a = a'_{A1}$ may be assumed, the distribution of the valence electrons on the components prevents a NaTl type from becoming stable.
NaTl(F 2.2, SR 3.19) $a = a_W(2) = a'_{A1}(2)$; the structure may be described as a Si-type of Tl atoms filled with Na atoms.
Tl_7Sb_2(B 21.6, SR 3.175) $a = a_W(3)$, binding perhaps related to that of Tl.h.
$Li_{22}Pb_5$(F 88.20, SR 22.158) $a = a_W(6)$.

Vacancy variants with random distribution of the vacancies are a special kind of replacement variants of the W-family. As an example we consider $Ni_{1-x}Al_{1+x}$(C 1.1); for $x < 0$ a statistical replacement of Al by Ni was found and for $x > 0$ a statistical replacement of Ni by vacancies *(Bradley's partial occupation law)* Figure 16. Similar partial occupancy may be found in homotypic phases with other atomic compositions e.g. Cu_3Sb.h. Therefore partial occupation occurs in brass-like structures of the W-family when $N_V^{/A} > 1.5$; an explanation by means of atomic radii does not account for that observation. The number of valence electrons per cell during vacancy formation is constant in NiAl, in other examples for one atom dropping out of the cell one valence electron enters *(Norbury's rule)* and there are even phases in which for for one atom, more than one electron enters the cell (Fig. 16). This observation is not compatible with an explanation by the band model, which suggests that the Fermi sphere remains in contact with the Brillouin plane (110), since this would request a constant number of valence electrons in the cell. In any case the partial occupancy is a mechanism which allows the conservation of an energetically favorable number of electrons per cell, when the valence electron concentration increases. In the two-correlations model we can describe an energetically favorable correlation by a factorial commensurability of VEC and CEC. The factorial commensurability may be conserved by dropping out of the cell of atoms with low valence electron contribution. This dropping out changes the valence electron concentration without destroying the factorial commensurability between VEC and CEC. A change of $N_{P'}^{/C}$ is to be exprected when the commensurability of the electron correlations with the crystal cell changes. The phenomenon of partial occupancy is a general one; it is also found in structures of the diamond family. However in the CuZn-family the energy of vacancy formation is smaller than, for instance, in a closest-packing, as the nearest neighbors of a central atom are not nearest neighbors of one another; therefore in the case that the central atom drops out, the neighbors go nearer to one another, thus filling the vacancy to

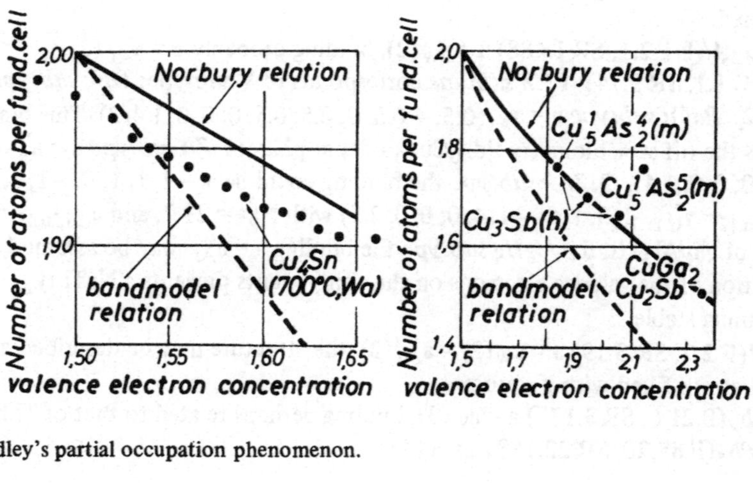

Fig. 16. Bradley's partial occupation phenomenon.

some extent. This possibility is not open in a closest-packing. When the number of vacancies per cell is sufficiently great, the vacancies will be ordered; these structures will be described below.

Shift of the homogenity range of phases homotypical to CuZn: Phases like Cu_3Sn or Cu_3Sb.h being homotypical to CuZn have a significantly higher valence electron concentration than $N_V^{/A} = 1.5$ and at the same time vacancies conforming to *Bradley's* low. At $N_V^{/A} = 1.5$ an expensive ionization of the $B^{4..5}$ atoms would be necessary because of the uniformity of the VEC. The ionization can be reduced by increasing the concentration of the B atoms; since the stability valence electron concentration is higher than 1.5 the CuZn-related phase has vacancies without being homogenous at a composition of full occupancy.

Ekman rule: Transition metals $T^{7...10}$ contribute zero electrons to the valence electron gas in brass-like phases. Electroneutrality forbids a negative VE contribution.

2. Substitutional Variants with Stronger Deformations

When a component of a phase has no filled d shell as outer core shell then the phase is not fully brass-like; the CEC will be different from the CEC in brass-like phases, introducing stronger deformations, and therefore the structures will be different.

We mention only two structure types of A–B phases, the binding of which has not yet been found.

Li_7Pb_2(H 7.2, SR 20.136) $a = a_{W.H}(1; 1; 3)$.

Li_8Pb_3(N 8.3, SR 20.138) $a = a_{W.H}(-1, 1, -1/3; 1, 1, 1/3; 0, 0, 3 + 2/3)$.

Most structure types of the subfamily of non-brass-like structures occur with T–B phases:

Zr_2Cu(U 2.1, SR 13.109) $a = a_W(1; 1; 3) = (3.3; 3.3; 11.3)$Å. Since the T component is in the majority, it may be that the CEC is not of the A2 type. A possibility for the binding is $a = a'_{A2}(1, -1, 0; 1, 1, 0; 0, 0, 5) = a''_{A0}(2, -2, 0; 2, 2, 0; 0, 0, 10)$, once more we have a factorial VC commensurability. It is clear that the CEC in the neighborhood of Cu is an open problem.

Nb_3Au_2(U 3.2, SR 24.121) $a = a_W(1; 1; 5) = (3.38; 3.38; 15.2)$Å $= a'_{A0}(2; 2; 9) = a''_{A0}(2, -2, 0; 2, 2, 0; 0, 0, 12.7)$.

CuAu(T 1.1, SR 1.505), the type occurs in TiRh with the two-fold primitive cell $a_{TiRh} = (4.17; 4.17; 3.35)$Å $= a'_{A1}(1, -1, 0; 1, 1, 0; 0, 0, 1.13) = a''_{A2}(2, -2, 0; 2, 2, 0; 0, 0, 2.25)$.

VIr.α(Q 2.2, SR 30.64) $a = a_W(2; 2; 1) = (5.791; 6.756; 2.796)$Å $= a'_{A1}(2; 2.33; \approx 1) = a''_{A2}(4; 4.66; \approx 2)$.

Ti_2Pd_3(Q 4.6, SR 33.126) $a = \tilde{a}_W(1, 0, -1; 0, 5, 0; 1, 0, 1) = (4.61; 14.33; 4.64)$Å, chemical stacking $(BTBBT)^2$. Binding $a = a'_{A0}(1, -2, 0; 2, 1, 0; 0, 0, 7) = a''_{A2}(3, -1, 0; 1, 3, 0; 0, 0, 10)$.

Ti_3Pd_5(T 3.5, SR 33.127) $a = \tilde{a}_W(1; 1; 4) = (3.26; 3.26; 11.44)$Å, stacking BTBBTBBT. Binding $a = a'_{A0}(0.5, -1.5, 0; 1.5, 0.5, 0; 0, 0, 5.5) = a''_{A2}(2, -1, 0; 1, 2, 0; 0, 0, 8)$.

$MoSi_2$(U 1.2, SR 1.740) a = $\tilde{a}_W(1; 1; 3)$. $TiPd_2$(SR 33.127) is isotypic: $a_{TiPd2-} = (3.24; 3.24; 8.48)$Å $= a'_{A0}(0.5, -1.5, 0; 1.5, 0.5, 0; 0, 0, 4) = a''_{A2}(2, -1, 0; 1, 2, 0; 0, 0, 6)$. In the subtype of $MoSi_2$ stacking variants occur.
TiCu(T 2.2, SR 15.69) a = $\tilde{a}_W(1; 1; 2) = (3.12; 3.12; 5.92)$Å $= a'_{A1.IF}(1, -0.5, 0; 0.5, 1, 0; 0, 0, 1.5) = a''_{A2}(2, -1, 0; 1, 2, 0; 0, 0, 4.25 \approx 4)$. The VEC gives a distance of 2.8 Å for the Cu-valence electrons and $N_{P'}^{/C} = 3.75$, this correlation should be filled in the neighborhood of the Ti atoms to an A2 correlation $a = a'_{A2}(1.5, 0.5, 0; -0.5, 1.5, 0; 0, 0, 3)$. The filling of a given correlation in the neighborhood of certain components has been called a composed VEC (*Schubert*, 1964), and should not be confused with coexisting correlations such as a′ and a″. The commensurability element $K'_{33} = 1.5$ or = 3 may be the cause for the remarkable stacking sequence of layers normal to a_3: TTBB. The binding explains the substructure axial ratio or, better, the axial ratio hints at the assumed binding.
Ti_3Cu_4(U 3.4, SR 30.130) a = $\tilde{a}_W(1; 1; 7) = (3.13; 3.13; 19.94) = a'_{A1.IF}(1, -0.5, 0; 0.5, 1, 0; 0, 0, 5) = a''_{A2}(2, -1, 0; 1, 2, 0; 0, 0, 14)$. Chemical stacking $(TBBTTBB)^2$.
Ti_2Cu_3(T 4.6, ST 30.130) a = $\tilde{a}_W(1; 1; 5) = (3.13; 3.13; 13.95) = a'_{A1.IF}(1, -0.5, 0; 0.5, 1, 0; 0, 0, 3.5) = a''_{A2}(2, -1, 0; 1, 2, 0; 0, 0, 10)$. Chemical stacking BBTBBTTBBT.
VAu_2(Q 2.4, SR 27.226) is similar to Ti_2Pd_3.
V_4Zn_5(U 4.5, SR 29.92) a = $\tilde{a}_W(3; 3; 1) = (8.91; 8.91; 3.22)$Å $= a'_{A1.F}(2, -2, 0; 2, 2, 0; 0, 0, 1)$. It is assumed that the A1–VEC is filled to an A2–VEC in the neighborhood of the V, so that 32 more places are available for $N_V^{/C} = 60$ "valence electrons".

3. Variants with Ordered Vacancies and Weak Displacements

Many variants may be derived from the W type by inserting vacancies in a particular order. It is convenient to arrange these types following the subtypes of the W type and following increasing number of non-occupied places per substructure place (*non-occupancy* $N_N^{/P}$). In the Na subtype no vacancy variants have been found, but in the Be.h subtype the following structure has been observed:
Be_9B_2(T 9.2, SR 24.58 27.67) a = $\tilde{a}_W(1, -1, 0; 1, 1, 0; 0, 0, 3) = (3.369; 3.369; 7.050)$Å $= a'_{A1}(1, -1, 0; 1, 1, 0; 0, 0, 3)$. The atomic vacancies make possible the full occupation of the VEC.

Most of the vacancy variants belong to the CuZn subtype:
Cu_5Zn_8(B 10.16, SR 33.72) with a = $\tilde{a}_W(3)$ has $N_N^{/P} = 2/54 = 0.037$, the famous gamma brass. Several variants of this structure have been described: *Cu_5Cd_8*(B 10.16, SR 33.72) has a somewhat different distribution of the components; *Fe_3Zn_{10}*(B 6.20, SR 1.497) and *Cu_9Al_4*(C 36.16, SR 3.57) have a different composition; *$Cu_4Sn.h$* (*Arnberg et al.*, 1976) has a cell a = $\tilde{a}_W(6)$ since the Sn atoms are arranged according to a Fe_3Si type, while the vacancies obey a Cu_5Zn_8 type of ordering. Several distorted structures have been found, for instance *Cr_5Al_8*(R 10.16, SR 5.11). It is very probable that in these structures a binding of the kind $a'_{A2} = a''_{A2}(2)$ is valid, where

the AV commensurability is twinned; low-temperature modifications of phases with CuZn-type of structure allow to infer the untwinned AV commensurability (e.g., in AuCd.r).

Ni_3Ga_4(B 24.32, SR 34.137) $a = \tilde{a}_W(4)$ $N_N^{/P} = 8/64 = 0.125$, has vacancy chains similar to Ni_2Al_3, forming a three-dimensional tissue. The binding may be $a = a'_{A0}(4, -4, 0; 4, 4, 0; 0, 0, 6) = a''_{C11}(8; 8; 10)$ or it may be related to the binding of Ni_2Al_3.

Ni_2Al_3(H 2.3, SR 5.10 5.67) $a = \tilde{a}_W(1, -1, 1; -1, 0, 1; 0, 1, 1) = (4.04; 4.04; 4.90)$ Å $N_N^{/P} = 0.17$. The binding may be $a = a'_{AH}(1; 2; 2.43) = a''_{C11.H}(2, 2, 0; -2, 4, 0; 0, 0, 2.43)$ where AH is a hexagonal primitive lattice and C11.H the pseudohexagonally described C11 cell with two translation points per cell. The value $N_{P''}^{/C} = 58$ applies also to the isotype Ni_2Ga_3, but it cannot apply to NiGa due to lack of CEC places.

Pt_2Al_3(H 4.6, SR 29.12), the structure is erroneous (*Chattopadhyay* and *Schubert*, 1975).

$Ni_{1.2}Cu_{4.6}Al_8$(R 6.8, SR 20.13) $a = \tilde{a}_{Ni2Al3}(1; 1; 14)$. Further variants of this kind are described in SR 21.9.

$PtHg_2$(T 1.2, SR 17.226) $a = \tilde{a}_W(1, -1, 0; 1, 1, 0; 0, 0, 1)$ $N_N^{/P} = 1/4 = 0.25$. The binding is $a = a'_{A2}(1, -1, 0; 1, 1, 0; 0, 0, 1) = a''_{A2}(2, -2, 0; 2, 2, 0; 0, 0, 2)$ with $N_{P''}^{/A} = 10.6$; this low $N_{P''}^{/A}$ excludes the occurrence of the structure for "$PtZn_2$"; or "$PtCd_2$"; cf also $NiHg_4$.

CaF_2(F 1.2, SR 1.148) $a = \tilde{a}_W(2)$ $N_N^{/P} = 4/16 = 0.25$. A brass-like representative is $AuGa_2$ with the binding $a = a'_{A1}(2) = a''_{A2}(4)$ exactly obeying Norbury's rule; the representative $PtSn_2$ does not follow Norbury's rule since the electrons are more concentrated around The Sn atoms. The same is valid for Mg_2Sn. The *$Cu_2Se.h$*(SR 4.113) subtype shows a statistical distribution of the Cu over different interstices of the ordered Se partial structure. The phase is not stable at lower temperatures because the Se valence electrons are then no longer in an A1 correlation. The *CaF_2* subtype (SR 1.185 has perhaps the binding $a = a'_{A2}(2) = a''_{A2}(4)$. Further variants are discussed by *Hellner* (1976). Substitutional variants of the CaF_2 type:

AgMgAs(F 1.1.1, SR 8.28) with As in the Ca sites; $a = a'_{A1}(2) = a''_{A2}(4)$.

CuMgAs(F 1.1.1, SR 8.28) with Cu in the Ca sites.

Li_3AlN_2(B 48.16.32, SR 11.20) $a = \tilde{a}_{CaF2}(2)$, N on Ca sites.

Li_7MnN_4(C 56.8.32, SR 23.175) $a = \tilde{a}_{CaF2}(2)$.

Vacancy variants of the CaF_2 type:

Zn_3P_2(T24.16, SR 3.51) $a = \tilde{a}_{CaF2}(1, -1, 0; 1, 1, 0; 0, 0, 2)$ $N_N^{/P} = 24/64 = 0.375$ (needs confirmation).

Cd_3As_2(U 48.32, SR 33.25) $a = \tilde{a}_{CaF2}(2; 2; 4)$ $N_N^{/P} = 0.375$ (formerly considered Zn_3P_2 type).

Mg_3P_2(B 24.16, SR 2.40) $a = \tilde{a}_{CaF2}(2)$ isotypical to Mn_2O_3(B 16.24, SR 2.38).

Further vacancy variants of the W-family:

Ru_3Sn_7(B 6.14, SR 11.136) $a = \tilde{a}_W(3) = a'_{A1}(3) = a''_{A2}(6)$ $N_N^{/P} = 14/54 = 0.26$. The cell contains more Sn than $PtSn_2$ in order to fill the d shell of Ru.

$NiHg_4$(B 1.4, SR 2.711 17.225) $a = \tilde{a}_W(2)$ $N_N^{/P} = 6/16 = 0.375$. The binding is $a = a'_{A2}(2) = a''_{A2}(4)$. The non-stability of "$NiHg_2$" conforms to the rule that lighter phases have higher $N_{P''}^{/A}$.

Mn_2Hg_5(T 4.10, SR 26.191) $a = \tilde{a}_W(3,1,0;-1,3,0;0,0,1) = a'_{A2}(3,1,0;-1,3,0;$ $0,0,1) = a''_{A2}2$ K'. The magnetic cell is $a_M = a(1;1;2)$. In Li_2Sn_5(Mn_2Hg_5 type, SR 34.99) we find $a = a'_{A1}(3,1,0;-1,3,0;0,0,1)$ and the Li-electrons are probably in the CEC.
Ti_3In_4(T 6.8, SR 30.14 report incorrect) is a substitutional variant of Mn_2Hg_5. The binding is similar to that of Li_2Sn_5.

In the following members of the W-family the $N_V^{/A}$ is so high that the metallic property vanishes.

VE-rich vacancy variants of the W-family are collected in the Si-family ("tetrahedral structures"). As a prototype of this family diamond is often used. Since the diamond structure is metastable at room temperature, we use the structure of Si instead. *Si*(F 2, SR 1.21) $a_{Si} = (5,43)$ Å $= \tilde{a}_W(2) = a'_{A1}(2) = a''_{A2}(4)$ $N_N^{/P} = 0.5$, In-binding; each atom has four nearest neighbors. A replacement variant of this structure is *ZnS*(F 1.1, SR 1.76) $a = \tilde{a}_{Si} = (5.43)$ Å. The structure becomes stable when the cation is sufficiently small compared with the anion which is in a closest packing. The members of this type have (as follows from the Si-binding) $N_V^{/A} = 4$ *(rule of Grimm/Sommerfeld)*. The value $N_V^{/A} = 4$ occurs in the compositions B^4, B^3B^5, B^2B^6, B^1B^7, $B_2^3B^6$, $B_3^3B^7$, $B^2B_2^5$, $B_3^2B_2^7$, $B^1B_3^5$, $B_2^1B_3^6$ and three-component and four-component compositions may be constructed by adding two formula symbols with $N_V^{/A} = 4$. However according to spatial correlation the number of electrons is related not to atoms but to lattice places. Introducing "0" for a vacant lattice place further formulas like OB_4^5, OB_2^6, $O_3B_4^7$ are conceivable (*Parthé*, 1964).

Following *ordering variants* have been observed:
$FeCuS_2$(U 2.2.4, SR 2.48) $a = \tilde{a}_{Si}(1;1;2)$, representatives with more evident electron count are $CuGaSe_2$(SR 17.19) or $ZnGeAs_2$(SR 22.42).
CuAsS(O 4.4.4, SR 30.18) $a = \tilde{a}_{Si}(1.5,0,-0.5;0,1,0;1.5,0,0.5)$.

Vacancy variants of the Si-family:
$\tilde{G}a_2S_3.r$(F(2.7)/4.1, SR 12.177) $a = \tilde{a}_{Si}$ $N_N^{/P} = 0.58$; from the composition we see that the Si binding is valid.
$HgJ_2.rot$(T 2.4, SR 1.177 1.189) $a = \tilde{a}_{Si}(0.5,0.5,0;-0.5,0.5,0;0,0,2)$, alternate Zn layers $(001)_{ZnS}$ are left free.
SnJ_4(C 8.32, SR 1.234) $a = \tilde{a}_{Si}(2)$, $N_N^{/P} = 0.72$. The position of the J is of the Cu type and the position of the Sn probably gives a minimum quadrupole energy of the Sn dumbbells in [111] direction.

Stacking variants of the substructure like ZnO(H 2.2, SR 1.119) are not considered here.

Even the noble gas structures belong in last consequence to the W-family.
Ar(F 1, SR 1.71) $a = \tilde{a}_{Cu} = \tilde{a}_{A0}(2)$. The VEC is occupied to $N_N^{/P'} = 0.5$, the CEC is no longer of influence.

An interesting phenomenon is to be observed in the Si-family when the anion has an electron contribution greater than four; in this case the A1-VEC will be filled in to an A0 correlation in the neighborhood of the anion:
SiS_2(P 2.4, SR 3.37) $a = \tilde{a}_{Si}(1;1;2) = (5.61;5.54;9.57)$ Å $= a'_{A1.F}(2;2;3.5) = a''_{A2}(4;$ $4;7)$, where A1.F designates a partly filled A1 correlation. The S occupy the position

a = $a_{Cu}(1;1;2)$, the Si are tetrahedrally surrounded by S and fill the site $f_{Si}(\delta.0.25.0 + \delta.0.75.0 + \delta.5.25.5 + \delta.5.75.5)$ giving the phase a fiber-like habit. In compounds of the mixture P–S the filling of the fundamental A1-VEC is so great that molecular structures result. Therefore even in molecular structures the electron spatial correlation is of importance.

4. *Vacancy Variants with Stronger Deformations and Displacements*

$Pt_2Si.r$(U 2.1, SR 24.120) $a = \tilde{a}_W(1,-1,0;1,1,0;0,0,2) = (3.93;3.93;5.92)$Å $N_N^{/P} = 0.25$. Binding $a = a'_{A2}(1,-1,0;1,1,0;0,0,2.13) = a''_{A2}(2,-2,0;2,2,0;0,0,4.25)$; the tetragonal deformation affords a greater $N_{P''}^{/A}$.
ZrH_2(U 1.2, SR 16.102) $a = \tilde{a}_W(1,-1,0;1,1,0;0,0,2) = (3.52;3.52;4.45)$Å, axial ratio is different from that of $Pt_2Si.r$.
Pd_5TlAs(T 5.1.1, SR 35.22) $a = \tilde{a}_W(1,-1,0;1,1,0;0,0,2) = (4.00;4.00;7.04)$Å. The binding is probably $a = a'_{A1}(1;1;1.75) = a''_{A2}(2,-2,0;2,2,0;0,0,5)$.
$FeSi_2.h$(T 1.2, SR 1.587) $a = \tilde{a}_W(1;1;2)$ $N_N^{/P} = 0.25$. A brass-like representative is $CuGa_2$(SR 38.79) with $|a_3|/|a_1| = 2.06$; the binding $a = a'_{A1}(1;1;2) = a''_{C11}(2;2;5)$ fulfills Norbury's rule; K''_{33} explains the ordering of vacancies in planes; the fact that $N_{P''}^{/A}(CuGa_2) = 13 > N_{P''}^{/A}(AuGa_2) = 11$ conforms to the rule that $N_{P''}^{/A}$ is smaller in heavier phases.
Cu_2Sb(T 4.2, SR 3.33) $a = \tilde{a}_W(1,-1,0;1,1,0;0,0,2) = (4.00;4.00;6.10)$Å $N_N^{/P} = 0.25$. The binding is $a = a'_{A1}(1,-1,0;1,1,0;0,0,2.1 \approx 2) = a''_{C11}(2,-2,0;2,2,0;0,0,5.3 \approx 5)$. Variants are:
PbFCl(T 2.2.2, SR 2.362) $a = \tilde{a}_W(1,-1,0;1,1,0;0,0,2)$, brass-like representative NiPdGe (*Wopersnow* and *Schubert*, 1977).
Mn_3As(O 12.4, SR 15.19) $a = \tilde{a}_W(1,-1,0;1,1,0;0,0,5) = (3.79;3.79;16.29)$Å.
Ti_2Bi(T 8.4, Mh.Ch.89.1958.154) $a = \tilde{a}_W(1,-1,0;1,1,0;0,0,4)$.

Another subfamily if the W-family is given by:
$CuAl_2$(U 2.4, SR 1.491) $a = \tilde{a}_W(2) = (6.06;6.06;4.89)$Å $= a'_{A1}(2) = a''_{C11}(4)$. As compared with $CuGa_2$ the K''_{33} is smaller since the $N_C^{/A}$ is smaller in $CuAl_2$ than in $CuGa_2$.

Variants of $CuAl_2$:
$CoGe_2$(Q 4.8, SR 11.96) $a = \tilde{a}_{CuAl2}(1;1;2)$.
$PdSn_2$(U 8.16, SR 20.168) $a = \tilde{a}_{CuAl2}(1;1;4)$.
$RhSn_2$(U 3.6, SR 20.177) $a = \tilde{a}_{CuAl2}(1;1;3)$.
$PdSn_3$(Q 4.12, SR 23.123) $a = \tilde{a}_{CuAl2}(1;1;3)$.
$PtPb_4$(T 2.8, SR 15.90) $a = \tilde{a}_{CuAl2}(1)$.
$PtSn_4$(Q 2.8, SR 13.116) $a = \tilde{a}_{CuAl2}(1;1;2)$.
$CoGa_3$(T 4.12, SR 23.118) $a = \tilde{a}_{CuAl2}(1) = a'_{A1}(2;2;2.25) = a''_{C11}(4;4;5)$.
$PdGa_5$(U 2.10, SR 23.121) $a = \tilde{a}_{CuAl2}(1;1;2) = a'_{A2}(2,-2,0;2,2,0;0,0,4.4 \approx 4) = a''_{C11}(4;4;7.6 \approx 8)$.

Co_2Al_9(M 4.18, SR 11.8) a = $\tilde{a}_W$ (4, 0, 0; 0, 2, 0; 0.25, 0, 2) = (8.53, 0, 0; 0, 6.29, 0; –.71, 0, 6.21) = $a'_{A1.I}$(3 ≈ 3.25; 3; 3) = a''_{C11}(7 ≈ 6.5, 0, 0; 0, 3, – 3; 0, 3, 3); this binding gives an energy gain because of commensurability of the core electrons. $FeCu_2Al_7$ (T 4.8.28, SR 13.9) a = a_{CuAl2}(1; 1; 3) = $a'_{A1.I}$(3; 3; 5) = a''_{A2}(3, – 3, 0; 3, 3, 0; 0, 0, 10). Evidently this is the In binding.

Mg_5Ga_2(P 10.4, SR 28.23) a = $\tilde{a}_W$(4; 2; 2); A brass-like representative is Cu_5As_2.m (JLCM 23.1971.231). a = (5.98; 11.58; 5.49) = a'_{A1}(2; 4; 2) = a''_{C11}(4; 8; 4.5); if As = As^4 is assumed Norbury's rule is valid. An analogous structure without vacansies is Ti_3Pt_5(SR 33.127).

5. Variants with Row Shear

In a W structure all atoms lying on a [111] line are in contact; since such an atom row may be fairly easily moved in its own direction *(row shear)* there are several structures which follow from a W structure by a special distribution of row shears and this mechanism is often connected with partial occupancy. It is clear that both mechanisms are caused by the binding. Structural variants of the W type with row shear are often hexagonal or pseudohexagonal; it is therefore appropriate to use for comparison the cell $a_{W.H}$: = a_W(1, – 1, 0.5; 0, 1, 0.5; – 1, 0, 0.5) which contains three atoms in a fundamental cell. It is also appropriate to classify the structures according to their non-occupancy $N_N^{/P}$ (= number of non-occupied places over number of places). Variants with full occupancy:

Ni₂In.h(H 4.2, SR 11.132) a = $\tilde{a}_{W.H}$(1; 1; 2), ideal axial ratio is $|a_3| / |a_1| = 1.225$. The rows 2/3, 1/3, z and 1/3, 2/3, z show a shear. The cell $a_{W.H}$ would also be compatible with the formula Ni_2In, but in this case In atoms would come into contact; the real structure avoids this, and the row shear makes the In site more regular. The results of the row shear may also be described as a falling together of two layers parallel to the hexagonal basal plane neighbored in the W structure. It is useful to connect two neighbored rows which have the support number 1 by a line; such lines show the superstructure mesh in the hexagonal basal plane. The binding of Ni_2In may be assumed as a = $a'_{C11.H}K''/2$ = $a''_{C11.H}$(3, 1, 0; – 1, 4, 0; 0, 0, 5/2) where K″ = commensurability of the CEC.

$Ag_{2-}Zn_{1+}.r$(H 6.3, SR 15.120) a = $\tilde{a}_{W.H}$(1, 1, 0; – 1, 2, 0; 0, 0, 1) = $a'_{A2.H}K''/2$ = $a''_{A2.H}$ (4; 4; 7/3); the commensurability element K''_{33} is wirtten in a manner describing the number of CEC layers per cell. The $N_V^{/C}$ = 14 explains the stoichiometric shift as compared with the ideal composition Ag_2Zn. The ideal axial ratio with respect to the substructure is $|a_3| / |a_1| = 0.35$.

Fe_2P(H 6.3, SR 16.95) a = $\tilde{a}_{W.H}$(1, 1, 0; – 1, 2, 0; 0, 0, 1) the binding for the representative Pd_2Ge is a = $a'_{A2.H}K''/2$ = $a''_{A2.H}$(3, 1, 0; – 1, 4, 0; 0, 0, 3); since the cores of Pd_2Ge are more different than of Ag_2Zn, the $N_{P''}^{/C}$ is greater.

AuCd.r(H 9.9, *Alasafi* and *Schubert*, 1977) $a = \tilde{a}_{W.H}(1,1,0;-1,2,0;0,0,2) = a'_{A2.H}K''/2 = a''_{A2.H}(4;4;14/3)$.
Ti₅Ga₄(H 10.8, SR 27.204) $a = \tilde{a}_{W.H}(1,1,0;-1,2,0;0,0,2)$.
Au₇Ga₂.h(H 21.6, SR 37.87) $a = \tilde{a}_{W.H}(1,1,0;-1,2,0;0,0,3) = a'_{C11.H}K''/2 = a''_{C11.H}(6;6;4)$.
Pd₇Zn₃Ga₄(R 7.7, SR 39.61) $a = \tilde{a}_{W.H}(2,1,0;-1,3,0;0,0,2) = a'_{A2.H}K''/2 = a''_{A2.H}(6;6;13/3)$.
PdAl(R 13.13, *Matkovic* and *Schubert*, 1977) $a = a_{W.H}(3,1,0;-1,4,0;0,0,2) = a'_{A2.H}K''/2 = a''_{A2.H}(7,2,0;-2,9,0;0,0,13.5/3)$.

Structures with non-occupancy $0 < N_N^{/P} < 0.1$ (array following size of superstructure cell):

Cu₁₀Sn₃.m(H 20.6, SR 37.145) $a = \tilde{a}_{W.H}(1,1,0;-1,2,0;0,0,3)$	$N_N^{/P} = .037$
Pd₂₅Ge₉(H 25.9, JLCM 41.1975.97) $a = \tilde{a}_{W.H}(1,1,0;-1,2,0;0,0,4)$	.055
Pd₈Sb₃(R 32.12, JLCM 48.1976.79) $a = \tilde{a}_{W.H}(1,1,0;-1,2,0;0,0,15)$	.022
Ni₃₁Si₁₂(H 31.12, SR 37.114) $a = \tilde{a}_{W.H}(1,1,0;-1,2,0;0,0,5)$	.044
Pd₃₁Sb₁₂(R 31.12, JLCM 1977) $a = \tilde{a}_{W.H}(1.1,0;-1,2,0;0,0,15)$	.044
Pd₅Sb₂(H 30.12, SR 35.14) $a = \tilde{a}_{W.H}(1,1,0;-1,2,0;0,0,5)$	.089
Ti₆Sn₅(H 12.10, SR 30.169) $a = \tilde{a}_{W.H}(2;2;2)$	.083
Ni₄GaGe₂(H 39.27, JLCM 46.1976.139) $a = \tilde{a}_{W.H}(2;2;6)$	.083
Au₇In₃(H 42.18, JLCM 41.1975.33) $a = \tilde{a}_{W.H}(2.1,0;-1,3,0;0,0,3)$	.048
Pd₁₂Ga₂Ge₅(H 12.7, ZM 66.1975.353) $a = \tilde{a}_{W.H}(2,1,0;-1,3,0;0,0,1)$	.095
Pd₂₀Sb₇(R 20.7, JLCM 51.1977.35) $a = \tilde{a}_{W.H}(2,1,0;-1,3,0;0,0,4)$	.036

Structures with non-occupancy $0.1 < N_N^{/P} < 0.3$

Pt₂Sn₃(H 4.6, SR 11.177) $a = \tilde{a}_{W.H}(1;1;4)$	.16
Mn₅Si₃(H 10.6, SR 4.24) $a = \tilde{a}_{W.H}(1,1,0;-1,2,0;0,0,2)$	.125
Cu₃P.r(H 18.6, SR 38.86) $a = \tilde{a}_{W.H}(1,1,0;-1,2,0;0,0,3)$	.11
FeSi(C 4.4, SR 2.13) $a = \tilde{a}_{W.H}(1,1,0;-1,2,0;0,0,3)$	.11
Rh₂₀Si₁₃(H 20.13, SR 30.88) $a = \tilde{a}_{W.H}(3,1,0;-1,4,0;0,0,1)$	.15

A special subfamily of the W-family showing row shear and vacancies is the NiAs type with the prototype:
NiAs(H 2.2, SR 1.84) $a = \tilde{a}_{W.H}(1;1;2)$. The vacancies (non-occupancy positions N) as compared with the Ni_2In structure are g_N: $= f_N^*(\delta.67.33.25 + \delta.33.67.75)$ while the atom sites are $f_{Ni}^*(\delta.0.0.0 + \delta.0.0.5) + f_{As}^*(\delta.67.33.75 + \delta.33.67.25)$. The site of the atoms corresponding to As has to be considered as an indication of membership to the NiAs-family. Many members of the NiAs-family contain disordered vacancies, however here we shall consider only structural types with ordered vacancies. According to the mechanism of vacancy formation the non-occupancy increases with increase of $N_V^{/A}$. The ordering of the vacancies often lowers the symmetry. The binding is not exactly the same in all members of the NiAs-family. As an example we have $a_{NiSb} = (3.96;3.96;5.16)\text{Å} = a'_{C11.H}(2;2;3/2) = a''_{C11.H}(2,2,0;-2,4,0;0,0,5.2\approx 5)$; this CEC gives a possibility to explain the remarkable fact that in several vacancy variants of NiAs only every second T-atom layer parallel to the basal plane contains vacancies.

Variants with non-occupancy (relative to Ni_2In) $N_N^{/P} < 0.33$:

Ni_5Ge_3.r(N 20.12, SR 37.89) a = $\tilde{a}_{W.H}(3, -1, 1; 3, 1, 1; 0, 0, 2)$ $N_N^{/P}$ = .11
Ni_3Sn_2.r(O 12.8, SR 32.107) a = $\tilde{a}_{W.H}(-1, 2, 0; 1, 2, 0; 0, 0, 2)$ .16
CuAl.r(N 5.5, SR 38.4) a = $\tilde{a}_{W.H}(-1, 1, -1; 1, 1, 1; 4, 0, 0)$ .17
$Ni_{19}Ge_{12}$(N 38.24, SR 37.90) a = $\tilde{a}_{W.H}(3, -1, 0; 3, 1, 0; 0, 0, 4)$ .14
$Ni_{13}Ga_9$(N 26.18, SR 34.137) a = $\tilde{a}_{W.H}(-2, 2, -1; 2, 2, 1; 0, 0, 2)$ .08
$Pd_{13}Tl_9$(H 13.9, SR 34.137) a = $\tilde{a}_{W.H}(2; 2; 2)$ .04
Ni_5P_4(H 20.16, SR 30.70) a = $\tilde{a}_{W.H}(2; 2; 4)$ .25
Ni_4GaGe_2(H 39.27, JLCM 46.1976.139) a = $a_{W.H}(2; 2; 6)$ .08

Structures with non-occupancy $0.33 \leqslant N_N^{/P}$ (*Kjekshus* and *Pearson*, 1964)

FeS.r(H 12.12, SR 20.129) a = $a_{W.H}(1, 1, 0; -1, 2, 0; 0, 0, 4)$
Fe_3Se_4(N 3.4, SR 20.127) $a_I = a_{W.H}(1, 1, 0; -1, 1, 0; 0, 0, 4)$ where a_I is an inner-centered cell which is better adapted to the NiAs cell than the N 3.4 cell, the a_{I3} axis corresponding to the a_3(NiAs) axis. The Fe rows parallel a_{I3} may have the foot points (A, B) = (0.0.0, .5.5.0); then we write the symbol 0 or A or B if a Fe layer parallel $(001)_I$ has no vacancy, a vacancy on the A row or a vacancy on the B row. Writing for a Se layer simply "," we obtain the stacking symbol of Fe_3Se_4 as A, 0, B, 0, .
Cr_5Se_6(10.12, SR 21.95) a = $a_{W.H}(1, 1, 0; -1, 2, 0; 0, 0, 4)$. For (A, B, C) = (.0.0.0, .67.33.0, .33.67.0) we obtain the stacking B, 0, C, 0, .
$Cr_{2+}S_3$(H 8.12, SR 21.96) a = $a_{W.H}(1, 1, 0; -1, 2, 0; 0, 0, 4)$, stacking A + B, 0, B + C, 0,.
Cr_2S_3(R 4.6, SR 21.96) a = $a_{W.H}(1, 1, 0; -1, 2, 0; 0, 0, 6)$, stacking A + B, 0, B + C, 0, C + A, 0, .
Fe_7Se_8.h(H 21.24, SR 20.126) a = $a_{W.H}(2; 2; 6)$ with A, B, C, D = (.0.0.0, .5.0.0, .0.5.0, .5.5.0) we get the stacking A, 0, B, 0, C, 0,.
Fe_7Se_8.r(Z 56.64, SR 23.167) a = $a_{W.H}(2, 2, 0; -2, 2, 0; 0, 0, 8)$
Fe_7S_8(N 14.16, SR 17.444) $a_F = a_{W.H}(2, 2, 0; -2, 2, 0; 0, 0, 8)$ = (all face-centered cell), stacking A, 0, B, 0, C, 0, D, 0, .
Cr_7Se_8(N 7.8, SR 26.116) $a_F = a_{W.H}(2, 2, 0; -2, 2, 0; 0, 0, 4)$
CdI_2(H 1.2, SR 1.213 1.780) a = $a_{W.H}(1; 1; 2)$.

IV. Concluding Remarks

The preceding expositions show that many structural variants are stabilized in the Cu- and W-family. These variants are derived from the fundamental structures by the few mechanisms of substitution, vacancy formation, deformation, shear, and displacement. Most of the mechanisms are explained only be the electronic spatial correlation which provides a link between chemical content of the crystal and its geometrical measures. Substituting (in an experiment of thought) gradually one component of a binary mixture in some structure gives a continuous change of the number of valence electrons per cell; however only at a certain valence electron concentration do energetically preferable commensurabilities to the lattice become possible. Therefore the experiment of thought affords an argument for the stability of some structure at a certain composition, in other words the correlations model is a valence model for metallic phases. Since the times where it was understood that chemical elements may form metallic compounds, the problem of valence in alloy phases has only partially disclosed the essential parameters. In Berzelius' electrostatic theory the atomic charge was considered. In Mott's band model theory the influence of the lattice on the momentum of the electrons was considered. In Laves' arguments for the existence of phases like Cu_2Mg the atomic radius was considered. However, the microscopic theory teaches us that only the consideration of the spatial correlations builds the necessary bridge between the electronic structure of the components and the geometric structure of the compound. Therefore only consideration of the spatial correlation, which comprehends also electric charges and atomic radii, gives the complete tools to construct a systematics of metallic phases, a natural system which reflects the energetic influences favoring certain compositions to become compounds and thermodynamic phases. The preceding expositions prove that an analysis toward the realized correlations is possible, as an analysis of atomic charges or atomic radii has been possible in the past. Since the models of atomic charges and of atomic radii have influenced the theory of chemical binding, it cannot be doubted that later a more elaborate theory will make use of the present empirically found concepts and statements.

Acknowledgements. This paper contains part of a series of lectures given during a stay in the Physics Department of Assiut University, Egypt, in October 1976. The author wishes to express many thanks to Prof. Dr. Ghobrial Abdel Macieh and his colleagues for this unforgettable stay in Egypt.

Abbreviations

AC = atomic spatial correlation
CEC = core electron spatial correlation
$D_s = 1/L$ = shear density = 1/shear length
$N_C^{/A}$ = number of core electrons per atom
$N_{P'}^{/A}$ = number of valence electron places per atom
$N_{P''}^{/A}$ = number of core electron places per atom
$N_V^{/A}$ = number of valence electrons per atom
$N_V^{/C}$ = number of valence electrons per cell
VEC = valence electron spatial correlation
VEP = valence electron places

Symbols for correlation types, used also as matrix indices:

A0 = cubic primitive translation lattice
A1 = cubic face-centered lattice
A2 = cubic body-centered lattice
AH = hexagonal primitive lattice with equal distances
C11 = tetragonal body-centered lattice with hexagonal (110) plane
$a^i_{A0.H}$ = hexagonal three-fold primitive cell of A0 lattice described in coordinates i; the i is mostly dropped.
$a^i_{A1.I}$ = innercentered cell of A1 (two-fold primitive)
$a_{A1.F}$ = an A1 cell which is filled to A0 or A2 in certain regions of the crystal cell.

The structure type description contains a character for Laue group and numbers counting the atoms of the components in the primitive cell. The characters are

C = cubic primitive
B = cubic body-centered
F = cubic face-centered
T = tetragonal primitive
U = tetragonal body-centered
H = hexagonal primitive
R = hexagonal rombohedral
O = orthorhombic primitive
P = orthorhombic body-centered
Q = orthohombic one-face centered
S = orthorhombic all-face centered
M = monoclinic primitive
N = monoclinic C-face centered
Z = triclinic

Designation of homologous classes:

A^1 = alcali,
A^2 = earth alcali,
$T^3 ... T^{10}$ = homologous classes of transition elements,
$B^1 ... B^7$ = classes of Cu ... Br,
B^8 = noble gases.

The numbers must be written as high indexes to have the low index place free for compound designation.

References

Alasafi, K.M., Schubert, K.: 1977, J. Less-Comm. Met. submitted.
Alasafi, K.M., Schubert, K.: 1977, to be published.
Arnberg, L., Johnson, A., Westman, S.: 1976, *Acta Chem. Scand.* **A30**, 187.
Bopp, F.: 1961. In: W. Heisenberg und die Physik unserer Zeit, Braunschweig, Vieweg.
Chattopadhyay, T., Schubert, K.: 1975, J. Less-Comm. Met. **41**, 19.
Chattopadhyay, T., Schubert, K.: 1976, J. Less-Comm. Met. **45**, 79.
Fujiwara, K.: 1975, J. Phys. Soc. Japan **12**, 7.
Hellner, E.: 1976, Z. anorg. allg. Chem. **421**, 49.
Hurley, A. C.: 1976, Electron correlation in small molecules. London: Acad. Press.
Kaufman, L., Bernstein, H.: 1970, Computer calculations of Phase Diagrams. New York: Academic-Press.
Khalaff, K., Schubert, K.: 1974, J. Less-Comm. Met. **37**, 129.
Kjekshus, A., Pearson, W.B.: 1964, Progr. Solid State Chem. **1**, 83.
Matković, T., Schubert, K.: 1977, J. Less-Comm. Met. submitted.
Parthé, E.: 1964, Crystal Chemistry of Tetrahedral Structures. New York: Gordon Breach.
Pearson, W.B.: 1972, The Crystal Chemistry and Physics of Metals and Alloys. New York: Wiley.
Sato, H., Toth, R.S.: 1965, Metallurgy Soc. Conf., Vol. **29**.
Schubert, K.: 1964, Kristallstrukturen zweikomponentiger Phasen. Berlin: Springer.
Schubert, K.: 1974, Acta Cryst. **B30**, 193.
Schubert, K.: 1975, J. Less-Comm-Met. **43**, 97.
Schubert, K.: 1976, J. Less-Comm. Met. **44**, 23.
Structure Reports, Vol. 1...7. Leipzig: Akad. Verl. Gesellschaft (1931...1943).
Structure Reports, Vol. 8 ... , published for the IUCr, Utrecht, Bohn, Scheltema, Holkema (1956...).
Wigner, E., Seitz, F.: 1933, Phys. Rev. **43**, 804.
Wopersnow, W., Schubert, K.: 1977, J. Less-Comm. Met. **52** (1977) 1.

Rare-Earth – Oxygen Bonding in the $LnMO_4$ Xenotime Structure: Spectroscopic Investigation and Comparative Study of Ligand Field Models

C. Linares, A. Louat and M. Blanchard

Physico-Chimie des Matériaux Luminescents, E.R. N° 10, 43, Bd. du 11 Novembre 1918, 69621 Villeurbanne – France

Table of Contents

The aim of this paper is to investigate the character of rare-earth-oxygen bonding, from a systematic study of the fluorescence spectra of the Europium trivalent ion incorporated in some isostructural compounds $LnMO_4$ ($Ln = Gd^{3+}, Y^{3+}, Lu^{3+}$ and $M = P^{5+}, As^{5+}, V^{5+}$).

Crystal field parameters and, in a ligand field hypothesis, intrinsic parameters relative to europium-oxygen bonding are determined.

The angular overlap model of Jørgensen is applied and also a more general model involving other effects than σ ones.

I. Structure

All the studied compounds have the tetragonal zircon structure (spatial group: D_{4h}^{19} (*1*)). The rare-earth ion is surrounded by a dodecahedron of 8 oxygens, 4 at a distance R_1 and 4 at a different distance R_2, so that the site symmetry is D_{2d} as can be seen in Fig. 1. Oz is the "c" axis (axis $\overline{4}$) and Ox is at 45° to the "a" axis (axis 2). The choice of these axes takes into account the character table of the D_{2d} group (*2*).

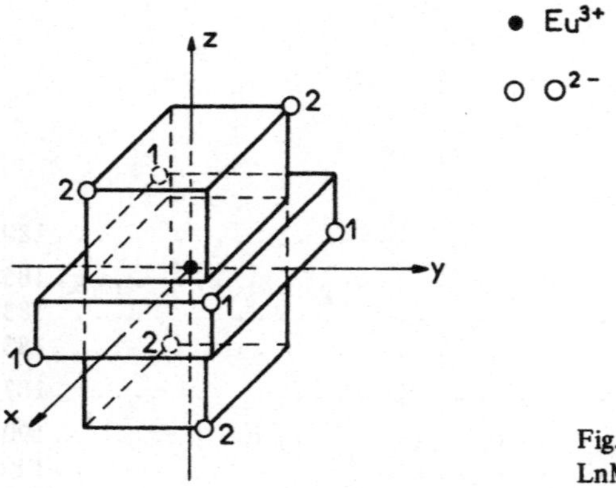

Fig. 1. The Eu^{3+} ion with its surrounding in $LnMO_4$.

The position of the farthest ions can be determined from Tables 1 and 2 (*3, 4, 5*). *Schwarz* (*3*) has given A and C values for all the compounds; accurate measurements of the oxygen positions in YVO_4, $LuAsO_4$ and $LuPO_4$ have been carried out by *Lohmüller et al.* (*4*). *Newman* and *Urban* (*6*) have assumed the MO_4 (M = P, As, V) complex to be rigid; the oxygen positions in the other compounds can then be extrapolated from the three studied compounds. Baglio and Sovers have demonstrated the accuracy of this method of extrapolation on the vanadates. This method has been applied here to $GdAsO_4$.

In Table 3 are given the 8 oxygen ligand coordinates which have been calculated from the various crystallographic data.

Table 1: Crystallographic data for $LnMO_4$.

Number of atoms per cell: 24

Coordinates x y z of atoms (origin at $\bar{4}2m$):

$$Ln^{III} \text{ rare-earth} \quad \begin{cases} 0, 0, 0 \\ 0, \frac{1}{2}, \frac{1}{4} \end{cases} + \begin{cases} 0, 0, 0 \\ \frac{1}{2}, \frac{1}{2}, \frac{1}{2} \end{cases} + h, k, l$$

$$M^V: \text{P, As, V} \quad \begin{cases} 0, 0, \frac{1}{2} \\ 0, \frac{1}{2}, \frac{3}{4} \end{cases} + \begin{cases} 0, 0, 0 \\ \frac{1}{2}, \frac{1}{2}, \frac{1}{2} \end{cases} + h, k, l$$

$$\text{oxygen} \quad \begin{cases} 0, u, v \\ 0, \bar{u}, v \\ u, 0, \bar{v} \\ \bar{u}, 0, \bar{v} \\ 0, \frac{1}{2} + u, \frac{1}{4} - v \\ 0, \frac{1}{2} - u, \frac{1}{4} - v \\ u, \frac{1}{2}, \frac{1}{4} + v \\ \bar{u}, \frac{1}{2}, \frac{1}{4} + v \end{cases} + \begin{cases} 0, 0, 0, \\ \frac{1}{2}, \frac{1}{2}, \frac{1}{2} \end{cases} + h, k, l$$

Transformation matrix

$$\begin{matrix} \frac{A\sqrt{2}}{2} & \frac{A\sqrt{2}}{2} & 0 \\ -\frac{A\sqrt{2}}{2} & \frac{A\sqrt{2}}{2} & 0 \\ 0 & 0 & C \end{matrix}$$

A, C: lattice parameters

Table 2. Lattice and positional parameters for $LnMO_4$.

Compounds	Lattice parameters		Positional parameters	
	$A_Å$	$C_Å$	u	v
YPO_4	6.888	6.021	0.1736	0.3407
$LuPO_4$	6.792	5.955	0.1760	0.3389
$GdAsO_4$	7.138	6.353	0.1788	0.3285
$YAsO_4$	7.044	6.248	0.1808	0.3278
$LuAsO_4$	6.949	6.227	0.1833	0.3239
$GdVO_4$	7.211	6.350	0.1832	0.3272
YVO_4	7.12	6.289	0.1852	0.3251
$LuVO_4$	7.026	6.231	0.188	0.3239

Table 3. Oxygen ion coordinates (R, θ, Φ).

Compounds	ligands	R_A	θ°	Φ°
YPO_4	2 oxygens	2.313	76.33	– 45, 135
	2 oxygens	2.313	103.67	45, – 135
	2 oxygens	2.374	30.22	45, – 135
	2 oxygens	2.374	149.78	– 45, 135
$LuPO_4$	2 oxygens	2.264	76.53	– 45, 135
	2 oxygens	2.264	103.47	45, – 135
	2 oxygens	2.346	30.95	45, – 135
	2 oxygens	2.346	149.05	– 45, 135
$GdAsO_4$	2 oxygens	2.346	77.72	– 45, 135
	2 oxygens	2.346	102.28	45, – 135
	2 oxygens	2.446	31.44	45, – 135
	2 oxygens	2.446	148.56	– 45, 135
$YAsO_4$	2 oxygens	2.300	77.80	– 45, 135
	2 oxygens	2.300	102.20	45, – 135
	2 oxygens	2.412	31.88	45, – 135
	2 oxygens	2.412	148.12	– 45, 135
$LuAsO_4$	2 oxygens	2.248	78.03	– 45, 135
	2 oxygens	2.248	101.97	45, – 135
	2 oxygens	2.385	32.27	45, – 135
	2 oxygens	2.385	147.73	– 45, 135
$GdVO_4$	2 oxygens	2.336	77.89	– 45, 135
	2 oxygens	2.336	102.11	45, – 135
	2 oxygens	2.462	32.45	45, – 135
	2 oxygens	2.462	147.55	– 45, 135
YVO_4	2 oxygens	2.291	78.10	– 45, 135
	2 oxygens	2.291	101.90	45, – 135
	2 oxygens	2.433	32.83	45, – 135
	2 oxygens	2.433	147.17	– 45, 135
$LuVO_4$	2 oxygens	2.242	78.18	– 45, 135
	2 oxygens	2.242	101.82	45, – 135
	2 oxygens	2.408	33.18	45, – 135
	2 oxygens	2.408	146.82	– 45, 135

II. Crystal Field Parameters

These parameters are determined from the optical spectra of Eu^{3+} in the different polycrystalline samples.

II.1. Excitation and Emission Spectra

Brecher et al. (*7, 8*) have only studied $YPO_4(Eu^{3+})$ and $YVO_4(Eu^{3+})$ as single crystals.

We prepared polycrystalline samples of the $LnMO_4$ series doped with 5% Eu^{3+} (Ln = Gd, Y, Lu and M = P, As, V) as has been shown in papers (*9*) and (*10*). $GdPO_4$ has not been studied because its structure is different.

The spectra of polycrystalline YPO_4 and YVO_4 are the same as those of the single crystals. Then, for the new compounds, the energy of the levels $^5D_{0,1,2}$ and $^7F_{0,1,2,3,4}$ has been determined (Table 4). The comparison with the levels found by *Brecher* was very valuable, especially for the arsenates, because only the intense lines are observed.

In Table 4, we see that the energy level positions in the arsenates and the vanadates are quite similar, those of the phosphates are more different. Also shown are the irreducible representations of D_{2d} appropriate to each level: they are in the same order for the arsenates and vanadates.

For any one rare-earth ion the $^5D_0$ energy level generally decreases from phosphates to arsenates and vanadates. It is well established that this decrease is connected with an increase in covalency. In this way, it will be noted that the arsenates behave as the phosphates. For each type of compounds, we can see that the $^5D_0$ energy level decreases from Gd to Y and Lu.

II.2. Crystal Field Parameters

With these experimental data, crystal field parameters B_n^m (exp) may be calculated.

Let us just recall that the crystal field at the rare-earth site may be written (*11*):

$$V = \sum_{n,m} B_n^m \, a_n \, O_n^m$$

B_n^m is the phenomenological crystal field parameter

$a_n = \alpha, \beta, \gamma$ for $n = 2, 4, 6$

O_n^m is the equivalent operator of Stevens.

Table 4. Trivalent europium energy levels (cm^{-1}).

Levels	represen-tation	PO_4		represen-tation	AsO_4			VO_4		
		YPO_4	$LuPO_4$		$GdAsO_4$	$YAsO_4$	$LuAsO_4$	$GdVO_4$	YVO_4	$LuVO_4$
$^7F_1$	E	339	349	A_2	370	353	334	350	334	317
	A_2	436	410	E	370	380	392	368	376	391
$^7F_2$	B_2	910	907	B_2	951	952	950	935	936	942
	A_1	1020	1005[b]	A_1	997[b]	998[b]	997[b]	993	985	980
	E	1065	1056	E	1040	1038	1033	1044	1039	1039
	B_1	1088	1104	B_1	1142[b]	1144[b]	1143[b]	1122	1116	1125
$^7F_3$	B_2	1832	1830	B_2	1874	1879	1860	1854	1855	1859
	E	1880	1881	A_2	1879[b]	1880[b]	1880[b]	1864	1873	1879
	A_2	1911	1911	E	1904[b]	1905[b]	1905[b]	1906	1904	1902
	B_1	1941	1925	B_1	1919[b]	1920[b]	1920[b]	1910	1903	1912
	E	1960	1950	E	1937	1942	1951	1954	1957	1966
$^7F_4$	B_1	2752	2752	B_1	2725[b]	2725[b]	2725[b]	2701	2700	2714
	E	2811	2806	E	2847	2844	2846	2818	2830	2825
	B_2	2848	2850	B_2	2872	2875	2879	2866	2868	2874
	A_1	2870[a]	2888	A_1	2875[b]	2875[b]	2875[b]	2884	2879	2868
	A_2	2911	2923	A_2	2875[b]	2875[b]	2875[b]	2918	2923	2903
	E	2997	2995	E	2980	2986	2992	2987	2988	3000
	A_1	3020[a]	3013	A_1	3035	3046	3039	3062	3063	3060
$^5D_0$	A_1	17211	17191	A_1	17214	17210	17200	17193	17183	17178
$^5D_1$	E	18965	18940	A_2	18967	18958	18946	18947	18932	18927
	A_2	18986	18955	E	18967	18964	18956	18949	18941	18941

[a] New levels, [b] presumed levels.

In our compounds, the Eu^{3+} symmetry site is D_2^d, so that the only parameters are B_2^0, B_4^0, B_4^4, B_6^0, and B_6^4. In this symmetry, and for the studied levels $^7F_{0,1,2,3,4}$ the energies corresponding to the irreducible representations B_1 and B_2 are always calculated from the wave functions $|2, 2\rangle \pm |2, -2\rangle$. The reference axes used are those of the character table, so that we may write without ambiguity:

$$|2, 2\rangle + |2, -2\rangle \text{ transforming as } B_1$$

and

$$|2, 2\rangle - |2, -2\rangle \text{ transforming as } B_2$$

for $^7F_2$ and $^7F_4$ levels while for $^7F_3$, B_1 and B_2 representations are inverted.

The Eu^{3+} crystal field parameters, reported in Table 5, have been computed with QCPE program No. 216 (*12*). This program permits us to find a set of parameters giving the weakest quadratic mean error between experimental and calculated levels, on condition that the error be increasing from $^7F_1$ to $^7F_4$ (the sublevels for small J being more reliable).

Table 5. Crystal field parameters (cm^{-1}).

	B_n^m	B_2^0	B_4^0	B_4^4	B_6^0	B_6^4
	$GdAsO_4$	– 11.9	21.2	– 631.6	– 53.4	+ 10
	$GdVO_4$	– 33	43.5	– 804	– 49.8	– 21
	YPO_4	150	39	– 733.5	– 46.5	– 100
Eu^{3+}	$YAsO_4$	– 61.1	25.7	– 738.5	– 51.4	– 9
	YVO_4	– 54.5	47.5	– 736.2	– 54.8	– 13
	$LuPO_4$	100	48.1	– 795.5	– 40.9	– 93
	$LuAsO_4$	– 105.8	25.8	– 769.8	– 55.3	+ 12
	$LuVO_4$	– 106	44.9	– 742.5	– 53.7	+ 3
Nd^{3+}	YVO_4 (13)	– 68	78	– 1071	– 73	+ 176
Er^{3+}	YVO_4 (14)	– 102.8	45.5	– 968	– 43	– 22
Tm^{3+}	YVO_4 (15)	– 88	42	– 870	– 38	+ 35

In Table 5, the parameters relative to YPO_4 are not those of Brecher[1]) because there is an error in his paper concerning the calculation of the E levels of $^7F_3$ and $^7F_4$; it was not in fact possible to determine a new set of parameters giving an acceptable agreement for $^7F_3$ and $^7F_4$ using the experimental levels of Brecher. So we have retained the energy values of $^7F_3$ and modified some sublevels of $^7F_4$, with the same polarizations. These levels are shown in Table 4.

For YVO_4, in order to have comparable results, new parameters have been calculated; these parameters are not very different from those of *Brecher*, and are within the limits of precision given by him. In fact, we found a set of parameters for which the mean quadratic error was slighter, for the $^7F_2$ level, than that of Brecher, the total error ($^7F_{1,2,3,4}$) not being increased.

In the second part of Table 5, parameters given in several publications are reported for other doping ions in the YVO_4 compound. It will be interesting to compare the results found for these different activators.

In Table 6 the different errors between experimental and calculated levels are reported for each of the ^7F multiplet sublevels and for all the multiplets. We note that the total error is greatest for the phosphates.

It is clear that some levels are uncertain and that $EuPO_4$ does not have the same structure as YPO_4 and $LuPO_4$ and therefore there is perhaps an additional perturbation. The error seems less important for the arsenates, but this is probably not significant: several levels are missing in the spectra of these compounds, which makes it easier to obtain an agreement.

[1]) Brecher's parameters (*7*): B_2^0 = 147; B_4^0 = 50; B_4^4 = – 740; B_6^0 = – 33; B_6^4 = – 392.

Table 6. Mean quadratic error (cm^{-1}).

ΔE	$^7F_1$	$^7F_2$	$^7F_3$	$^7F_4$	$^7F_{1,2,3,4}$
$GdAsO_4$	3.3	3.7	6.1	12.5	8.7
$GdVO_4$	0.6	5.4	9.3	15.1	10.8
YPO_4	4.2	6.3	14.1	23.1	14.7
$YAsO_4$	4.2	4.4	10.4	10.7	8.5
YVO_4	4.7	5	6.6	14.6	10
$LuPO_4$	1	5.5	10.2	21.9	14.3
$LuAsO_4$	2	1.8	1.8	6.8	4.3
$LuVO_4$	5.5	7.1	7.8	14.5	10.4

III. Point Charge Model

Though the electrostatic point charge model is a rough approximation, it may bring some interesting informations.

The electrostatic parameters were determined by Hutching's method (*16*). The complete calculation was reported in an earlier paper (*17*).

Results:

1. In Table 7, the electrostatic contribution of the ligands (B_n^m elec ligands) is compared to the contribution of all the ions of the crystals (B_n^m elec tot) and to the crystal field parameters (B_n^m exp) of Section II. The main results are the following:

B_2^0 – Let us compare:

a) B_2^0 (exp) with B_2^0 (elec tot) .

As is commonly the case, their values are very different; moreover, for the phosphates the signs are not the same.

b) B_2^0 (exp) with B_2^0 (elec ligands) .

These parameters have the same sign only for the phosphates.

This phosphate behaviour is the same as in $LaCl_3(Eu^{3+})$ (*17*), the most ionic of all the lanthanum compounds studied in this paper.

B_4^m – Let us compare:

a) B_4^0 (exp) with B_4^0 (elec tot) .

The signs of these two parameters are reversed for all the compounds. We never saw such a thing for the lanthanum series.

b) B_4^0 (exp) with B_4^0 (elec ligands) .

These parameters have the same signs and their values are close. From these facts, we may take it that the ligand electrostatic contributions form a great part of the B_4^0 (exp) and, moreover, that the rare-earth ion will not be very sensitive to the electrostatic field of the farthest ions. This screening phenomenon had been seen up to that time for the B_2^0 parameter only.

c) The three B_4^4 parameters (exp. elec. tot. and elec. ligands) .

They have nearly the same importance.

Table 7. Electrostatic and experimental parameters for Eu^{3+} in $LnMO_4$ (cm^{-1}).

Parameters	YPO_4	$LuPO_4$	$GdAsO_4$	$YAsO_4$	$LuAsO_4$	$GdVO_4$	YVO_4	$LuVO_4$
B_2^0 (elec ligands)	1371	1149	758	632	417	435	275	71
B_2^0 (elec tot)	− 1333	− 1598	− 1538	− 1795	− 2032	− 1868	− 2091	− 2380
B_2^0 (exp)	150	100	− 12	− 61	− 106	− 33	− 54.5	− 106
B_4^0 (elec ligands)	46	45.5	41	43	47	36.5	39	43
B_4^0 (elec tot)	− 21.5	− 23.5	− 10.5	− 13	− 9.5	− 16.5	− 16	− 15.5
B_4^0 (exp)	39	48	21	25.5	26	43.5	47.5	45
B_4^4 (elec ligands)	− 997	− 1116	− 944	− 1054	− 1184	− 979	− 1088	− 1205
B_4^4 (elec tot)	− 773	− 873	− 770	− 871	− 981	− 802	− 896	− 1011
B_4^4 (exp)	− 734	− 796	− 632	− 739	− 770	− 804	− 736	− 743
B_6^0 (elec ligands)	− 8.5	− 10	− 8	− 9.5	− 11	− 8.5	− 9.5	− 10.5
B_6^0 (elec tot)	− 14.5	− 16	− 12.7	− 14.5	− 16	− 12.5	− 14	− 15.5
B_6^0 (exp)	− 46.5	− 41	− 53.5	− 51.5	− 55.5	− 50	− 55	− 53.5
B_6^4 (elec ligands)	− 4.5	− 3	6	10	14.5	8	11.5	17
B_6^4 (elec tot)	+ 3.5	6	13.2	16.5	24.5	13.5	18.5	25
B_6^4 (exp)	− 100	− 92.5	10	− 9	12	− 21	− 13	3

B_6^m – Finally, let us compare:

a) B_6^0 (exp) with B_6^0 (elec tot) and B_6^0 (elec ligands).

The signs are the same in the three cases.

b) B_6^4 (elec tot) and B_6^4 (elec ligands).

These parameter values have nearly the same importance, the signs are reversed for the phosphates, but these numbers are about zero.

c) B_6^4 (exp) with B_6^4 (elec tot) and B_6^4 (elec ligands).

For the arsenates and vanadates the three numbers are nearly the same, though some signs sometimes differ; for the phosphates, B_6^4 (exp) is much greater than the two other parameters.

2. In Table 8, the ratios $B_n^m/B_n^{m'}$ (exp) and (elec ligands) are compared. It can be seen that these ratios are nearly the same, especially for the B_4^m ones.

Though the (elec tot) parameters are rather different from the B_n^m (exp), it is interesting to observe that a ligand field derived from an electrostatic one may be more valid for B_4^m than for B_6^m and then we may try to apply the superposition model.

Table 8. Calculated parameters ratios.

Ratios	YPO_4	$LuPO_4$	$GdAsO_4$	$YAsO_4$	$LuAsO_4$	$GdVO_4$	YVO_4	$LuVO_4$
B_4^4/B_4^0 (exp)	– 18.8	– 16.6	– 29.8	– 29	– 29.6	– 18.5	– 15.5	– 16.5
B_4^4/B_4^0 (elec ligands)	– 21.7	– 24.5	– 23	– 24.5	– 25.2	– 26.8	– 27.9	– 28
B_6^4/B_6^0 (exp)	2.15	2.26	0.19	0.17	– 0.22	'0.42	0.24	– 0.06
B_6^4/B_6^0 (elec ligands)	0.53	0.30	– 0.75	– 1.05	– 1.32	– 0.94	– 1.21	– 1.62

IV. Superposition Model

IV.1. Newman's Method (18).

The first assumption is that the total crystal field can be built from the separate contributions of each ligand around the rare-earth. The phenomenological parameters B_n^m are expressed in terms of intrinsic parameters $\bar{A}_n(R_j)$ depending only on the R_j distance of the ligand j from the lanthanide ion (*11*).

$$B_n^m = \sum_j K_n^m(j)\,\overline{A_n}(R_j)\,;$$

K_n^m is the coordination factor which depends only on the angular positions of the ligands.

In the case of a ligand environment not very much different from an electrostatic one it can be written (if R_j and R_k are not too different):

$$\overline{A_n}(R_j)/\overline{A_n}(R_k) = (R_k/R_j)^{t_n}\,;$$

t_n is a constant whose value is about $2n + 1$.

Two equations are then used:

$$B_n^0 = K_n^0(1)\,\overline{A_n}(R_1) + K_n^0(2)\,\overline{A_n}(R_1)\,(R_1/R_2)^{t_n}$$
$$B_n^4 = K_n^4(1)\,\overline{A_n}(R_1) + K_n^4(2)\,\overline{A_n}(R_1)\,(R_1/R_2)^{t_n}$$

$n = 4$ or 6. One t_n value must be found for which the $\bar{A}_n(R_1)$ or $\bar{A}_n(R_2)$ values will be the same from the two equations.

First, Newman's method was applied to yttrium phosphate, arsenate, and vanadate doped with Eu^{3+} (Table 9). The best results were obtained with $t_4 = t_6 = 8$.

- For $\bar{A}_4$, we see that Newman's method is right for YPO_4 and $YAsO_4$ but not for YVO_4.
- It can be also noted that it is not possible, with this method, to find one $\bar{A}_6$ parameter for any compound. *Newman* (*19*) had already noted this difficulty for the same compounds doped with Er^{3+}.
- Moreover we have the same difficulties for all the $\bar{A}_6$ and sometimes for the $\bar{A}_4$ of the other compounds doped with Eu^{3+} and for $YVO_4(Nd^{3+})$.

Thus, it has been necessary to modify Newman's method for application to our compounds in which crystal field is more different from an electrostatic one than usual, as has been seen before, especially for the $\bar{A}_6$.

Table 9. Intrinsic parameters for Eu^{3+} (cm^{-1}).

Intrinsic parameters ($t_n = 8$)		YPO_4	$YAsO_4$	YVO_4
$\bar{A}_4(R_1)$ from	↗ B_4^0	51.1	35.8	71.2
	↘ B_4^4	44.4	43.7	43.4
Mean value		47.8 ± 3.3	40 ± 3.9	57.3 ± 13.9
$\bar{A}_4(R_2)$ from	↗ B_4^0	41.5	24.5	44
	↘ B_4^4	36.1	29.9	26.8
Mean value		38.8 ± 2.7	27.2 ± 2.7	35.4 ± 8.6
$\bar{A}_6(R_1)$ from	↗ B_6^0	37.2	38.5	42.5
	↘ B_6^4	198	– 6	– 6.5
Mean value		?	?	?
$\bar{A}_6(R_2)$ from	↗ B_6^0	30.2	26.3	26.3
	↘ B_6^4	160.3	– 4.1	– 4
Mean value		?	?	?

IV.2. Generalization of the Method

We now write:

$$B_n^0 = K_n^0(1)\,\bar{A}_n(R_1) + K_n^0(2)\,\bar{A}_n(R_2)$$
$$B_n^4 = K_n^4(1)\,\bar{A}_n(R_1) + K_n^4(2)\,\bar{A}_n(R_2)$$

The power law which is adequate when the crystal field is sufficiently electrostatic is now canceled. The resolution of these two equations gives $\bar{A}_n(R_1)$ and $\bar{A}_n(R_2)$ directly. We have then chosen the new method to determine all the $\bar{A}_6$ and also some $\bar{A}_4$, using Newman's method only for the other $\bar{A}_4$ parameters.

The choice of the method will be justified on the one hand by some considerations on the covalency of the compounds and on the effect of host cation size on the bonding, and on the other hand by the use of these intrinsic parameters in an angular overlap calculation.

IV.3. Results

The results are listed in Table 10. When the $\bar{A}_4$ are found by Newman's method the error due to the difference between the two determinations from B_4^0 and B_4^4 is reported.

It can be seen, as we have said, that the generalized method is applied to all the $\bar{A}_6$ determinations and to the $\bar{A}_4$ determinations only for $GdVO_4$, YVO_4, and $LuVO_4(Eu^{3+})$ and for $YVO_4(Nd^{3+})$.

Let us explain this:

1. Among the intrinsic parameters, the $\bar{A}_6$ parameters are those in which the electrostatic contribution is the smallest, and, in this case, in order to calculate these parameters it is abviously normal to get rid of the power law.

Table 10. Intrinsic parameters (cm^{-1}).

				$GdAsO_4$		$GdVO_4$		
				(R_1)	(R_2)	(R_1)	(R_2)	
				$\bar{A}_4$ 32.8 ± 4.6	23.4 ± 3.2	50.2	1.1	
				$\bar{A}_6$ 34	28.8	28.7	27.2	
		YPO_4		$YAsO_4$		YVO_4		
		(R_1)	(R_2)	(R_1)	(R_2)	(R_1)	(R_2)	
Eu^{3+}	$\bar{A}_4$	47.8 ± 3.3	38.8 ± 2.7	40 ± 3.9	27.2 ± 2.7	47.3	− 14.4	10_1
	$\bar{A}_6$	22.5	30.4	30.7	27.5	31.8	28	
		$LuPO_4$		$LuAsO_4$		$LuVO_4$		
		(R_1)	(R_2)	(R_1)	(R_2)	(R_1)	(R_2)	
	$\bar{A}_4$	59 ± 11	44.3 ± 8.2	40.9 ± 4.7	25.4 ± 2.9	46.8	− 6.4	
	$\bar{A}_6$	18.8	25.5	34.4	28.4	33.0	26.8	

		Tm^{3+}		Er^{3+}		Eu^{3+}		Nd^{3+}		
		(R_1)	(R_2)	(R_1)	(R_2)	(R_1)	(R_2)	(R_1)	(R_2)	
YVO_4	$\bar{A}_4$	57.2 ± 5.9	35.3 ± 3.6	63 ± 6	38.9 ± 3.7	47.3	− 14.4	70.6	− 40.7	10_2
	$\bar{A}_6$	26.9	18.6	23.7	22.2	31.8	28	63.4	33.9	

	Lu^{3+}	Tm^{3+}	Er^{3+}	Y^{3+}	Gd^{3+}	Eu^{3+}	Nd^{3+}	
Ionic radius$_A$	0.99	1.04	1.04	1.06	1.11	1.13	1.15	10_3

2. The most covalent compounds among the three series of phosphates, arsenates, and vanadates are the vanadates, so it seems consistent to cancel the power law for these last compounds.

3. It can be also seen from Table 10_2 (referring to YVO_4 doped in various ways) that Newman's model is applied only to Er^{3+} and Tm^{3+}. In this table, the doping ions are set in order of increasing ionic radius (Table 10_3). The overlap is important for Eu^{3+} and still more for Nd^{3+}, and is why Newman's method is probably inadequate.

4. Justification by consideration of the $\langle r^n \rangle$ values: in the approximation of a crystal field with electrostatic features, B_n^m may be written:

$$B_n^m = A_n^m \langle r^n \rangle$$

A_n^m depends only on the coordinates of the ions; $\langle r^n \rangle$ are the mean values of the radial functions.

Therefore, in the same compound YVO_4 doped with different rare-earth ions, we must find exactly the same values for A_n^m. So Table 11 (a) is established using the

Table 11. (a) A_n^m values for YVO_4 (cm^{-1}). (b) calculated ratios.

	Tm^{3+}	Er^{3+}	Eu^{3+}	Nd^{3+}
$\langle r^2 \rangle$	0.64	0.67	0.83	1.00
A_2^0	138	154	66	68
$\langle r^4 \rangle$	1.04	1.13	1.75	2.40
A_4^0	41	41	27	32.5
A_4^4	– 837	– 858	– 421	– 446
$\langle r^6 \rangle$	3.54	5.1	7.4	12.40
A_6^0	– 11	– 8	– 7	– 6
A_6^4	10	– 4	– 2	14

(a)

$\frac{\langle r^4 \rangle (Er)}{\langle r^4 \rangle (Tm)}$	$\frac{B_4^0 (Er)}{B_4^0 (Tm)}$	$\frac{B_4^4 (Er)}{B_4^4 (Tm)}$
1.09	1.10	1.11
$\frac{\langle r^4 \rangle (Nd)}{\langle r^4 \rangle (Eu)}$	$\frac{B_4^0 (Nd)}{B_4^0 (Eu)}$	$\frac{B_4^4 (Nd)}{B_4^4 (Eu)}$
1.37	1.64	1.46

(b)

$\langle r^n \rangle$ values given by *Freeman* and *Watson* (*20*). It should be noticed that for the A_2^0 and A_4^m quite similar values are found for Tm^{3+} and Er^{3+} on the one hand and for Eu^{3+} and Nd^{3+} on the other, and also that the A_6^m differ a great deal from one compound to another (the crystal field is far from being an electrostatic one).

If we compare the experimental ratios to theoretical ratios given in Table 11(b), we observe that the values are about the same for Tm^{3+}/Er^{3+} and more scattered for Nd^{3+}/Eu^{3+}. The same calculations for B_6^m give bad results.

So we think that Newman's method is appropriate only for YVO_4: Er^{3+} and Tm^{3+} (for the $\overline{A}_4$ determinations).

IV.4. Conclusion

We have two groups of compounds:

– a group including the more ionic compounds $LnPO_4(Eu^{3+})$, $LnAsO_4(Eu^{3+})$ YVO_4 (Er^{3+} and Tm^{3+}); this group, for which the power law is retained, will be called for greater convenience, *group I.*

– a group including more covalent compounds: $LnVO_4(Eu^{3+})$ and $YVO_4(Nd^{3+})$, called *group II*; for this group the power law is suppressed.

V. Angular Overlap Model

V.1. Jørgensen Model (21)

In this model, we are interested in the relative extent of the σ antibonding influence of the ligands on f orbitals of the central rare-earth ion. The different calculations have been developed in paper (*11*). Let us briefly recall that the orbital f energy may be written:

$$E = \sigma^* \Xi^2$$

σ^* is the radial parameter, Ξ^2 the angular parameter which may be expressed by the angular coordinates of the ligands. E may be calculated with the crystal field parameters as shown in Table 12.

Table 12.

$$\langle \psi Z_3^0 | V | \psi Z_3^0 \rangle = \frac{8}{15} B_2^0 + \frac{16}{11} B_4^0 + \frac{1600}{429} B_6^0$$

$$\langle \psi Z_3^1 | V | \psi Z_3^1 \rangle = \frac{2}{5} B_2^0 + \frac{8}{33} B_4^0 - \frac{400}{143} B_6^0$$

$$\langle \psi Z_3^2 | V | \psi Z_3^2 \rangle = -\frac{56}{33} B_4^0 + \frac{160}{143} B_6^0 + \frac{8}{33} B_4^4 + \frac{160}{429} B_6^4$$

$$\langle \psi Z_3^{-2} | V | \psi Z_3^{-2} \rangle = -\frac{56}{33} B_4^0 + \frac{160}{143} B_6^0 - \frac{8}{33} B_4^4 - \frac{160}{429} B_6^4$$

$$\langle \psi Z_3^3 | V | \psi Z_3^3 \rangle = -\frac{2}{3} B_2^0 + \frac{8}{11} B_4^0 - \frac{80}{429} B_6^0$$

$$\langle \psi Z_3^3 | V | \psi Z_3^1 \rangle = -\frac{8\sqrt{15}}{165} B_4^4 + \frac{80\sqrt{15}}{1287} B_6^4$$

As an example, for the f orbital with the B_2 irreducible representation in D_{2d}, it can be written:

$$4\frac{\sqrt{7}}{2}\left[(5\cos^3\theta_1 - 3\cos\theta_1)^2\left(\frac{R_0}{R_1}\right)^8 + (5\cos^3\theta_2 - 3\cos\theta_2)^2\left(\frac{R_0}{R_2}\right)^8\right]\sigma_0^* =$$

$$= \frac{8}{15} B_2^0 + \frac{16}{11} B_4^0 + \frac{1600}{429} B_6^0$$

This equation gives a value of σ_0^* (for a mean distance $R_0 = (R_1 + R_2)/2$). The power 8 is given by *Jørgensen*. We have, then, five equations of this type in order to determine one value for σ_0^*.

Table 13. σ_0^* values (cm^{-1}).

Eu^{3+}	YPO_4				$LuPO_4$			
σ_0^*	38				38			
Representation (basis function)	E_{exp}	$\Xi^2 \sigma_0^*$	E_{calc}	$\lvert\Delta E\rvert$	E_{exp}	$\Xi^2 \sigma_0^*$	E_{calc}	$\lvert\Delta E\rvert$
$B_2(\psi z_3^0)$	− 36.7	− 2.3 σ_0^*	− 87.4	50.7	− 29.4	− 2.8 σ_0^*	− 106.4	77
$A_2(\psi z_3^2)$	− 333.4	− 8.1 σ_0^*	− 307.8	25.6	− 354.6	− 8.1 σ_0^*	− 307.8	46.8
$A_1(\psi z_3^{-2})$	97	2.3 σ_0^*	87.4	9.6	100.1	2.5 σ_0^*	95	6.1
E ψz_3^1	242.1	6.7 σ_0^*	254.6	12.5	229.7	6.4 σ_0^*	243.2	13.5
E ψz_3^3	− 105.6	2.7 σ_0^*	− 102.6	3	− 87.8	− 2.2 σ_0^*	− 83.6	4.2
ΔE_m (mean value)			23 cm^{-1}				35 cm^{-1}	

Eu^{3+}	$GdAsO_4$				$YAsO_4$				$LuAsO_4$			
σ_0^*	31				34				34			
	E_{exp}	$\Xi^2 \sigma_0^*$	E_{calc}	$\lvert\Delta E\rvert$	E_{exp}	$\Xi^2 \sigma_0^*$	E_{calc}	$\lvert\Delta E\rvert$	E_{exp}	$\Xi^2 \sigma_0^*$	E_{calc}	$\lvert\Delta E\rvert$
B_2	− 174.7	− 3.5 σ_0^*	− 108.5	66.2	− 186.9	− 3.7 σ_0^*	− 125.8	61.1	− 223.8	− 3.4 σ_0^*	− 115.6	108.2
A_2	− 245.1	− 8.1 σ_0^*	− 251.1	6	− 283.5	− 8.2 σ_0^*	− 278.8	4.7	− 287.9	− 8.3 σ_0^*	− 282.2	5.7
A_1	53.6	1.8 σ_0^*	55.8	2.2	81.3	1.9 σ_0^*	64.5	16.7	76.5	1.8 σ_0^*	61.2	15.3
E	225.8	7.2 σ_0^*	223.2	2.6	236.7	7.3 σ_0^*	248.2	11.5	256.5	7.8 σ_0^*	265.2	8.7
E	− 42.8	− 2.3 σ_0^*	− 71.3	28.5	− 42.1	− 2.4 σ_0^*	− 81.6	39.5	− 39	2.6 σ_0^*	− 88.4	49.4
ΔE_m			29				33				49	

Table 14. σ_0^* values.

Eu^{3+}	$GdVO_4$				YVO_4				$LuVO_4$			
σ_0^*	35				33				33			
	E_{exp}	$\Xi^2\sigma_0^*$	E_{calc}	$\lvert\Delta E\rvert$	E_{exp}	$\Xi^2\sigma_0^*$	E_{calc}	$\lvert\Delta E\rvert$	E_{exp}	$\Xi^2\sigma_0^*$	E_{calc}	$\lvert\Delta E\rvert$
B_2	– 140.1	$-4\sigma_0^*$	– 140	0.1	– 165.1	$-4.2\sigma_0^*$	– 138.6	26.6	– 191.5	$-4.4\sigma_0^*$	– 145.2	46.3
A_2	– 332.2	$-8.2\sigma_0^*$	– 287	45.2	– 325.4	$-8.3\sigma_0^*$	– 273.9	51.5	– 315.2	$-8.3\sigma_0^*$	– 273.9	41.3
A_1	73.1	$2.1\sigma_0^*$	73.5	0.4	41.1	$2\sigma_0^*$	66	24.9	42.6	$2\sigma_0^*$	66	23.4
E	250.3	$7.5\sigma_0^*$	262.5	12.2	251	$7.8\sigma_0^*$	257.4	6.4	256.2	$8.1\sigma_0^*$	267.3	11.1
E	– 50.8	$-2.4\sigma_0^*$	– 84	33.2	– 26.3	$2.5\sigma_0^*$	82.5	56.2	– 24.1	$2.7\sigma_0^*$	– 89.1	65
ΔE_m			25.5				38.5				43	

YVO_4	Nd^{3+}				Er^{3+}				Tm^{3+}			
σ_0^*	49				39				35			
	E_{exp}	$\Xi^2\sigma_0^*$	E_{calc}	$\lvert\Delta E\rvert$	E_{exp}	$\Xi^2\sigma_0^*$	E_{calc}	$\lvert\Delta E\rvert$	E_{exp}	$\Xi^2\sigma_0^*$	E_{calc}	$\lvert\Delta E\rvert$
B_2	– 195.1	$-4.2\sigma_0^*$	– 205.8	10.7	– 148.4	$-4.2\sigma_0^*$	– 169.8	15.4	– 127.6	$-4.2\sigma_0^*$	– 147	19.4
A_2	– 408	$-8.3\sigma_0^*$	– 406.7	1.3	– 369	$-8.3\sigma_0^*$	– 329.7	45.3	– 311.6	$-8.3\sigma_0^*$	– 290.5	21.1
A_1	20	$2\sigma_0^*$	98	78	116.7	$2\sigma_0^*$	78	38.7	84.1	$2\sigma_0^*$	70	14.1
E	402.6	$7.8\sigma_0^*$	382.2	20.4	276.9	$7.8\sigma_0^*$	304.2	27.3	260.8	$7.8\sigma_0^*$	273	12.2
E	91	$-2.5\sigma_0^*$	– 122.5	31.5	– 76.6	$-2.5\sigma_0^*$	– 97.5	20.9	– 83.2	$-2.5\sigma_0^*$	– 87.5	4.3
ΔE_m			35.9				29.6				13.9	

Kuse and *Jørgensen* (*22*) have already studied $ErPO_4$ and $ErVO_4$ as well as Eu^{3+} in YVO_4; we have completed this study. Table 12 was established from their publication with an adaptation to Stevens' B_n^m and to our axes.

The results are brought together in Tables 13 and 14. The σ_0^* value is determined by minimizing the difference $\Delta E = E_{calc} - E_{exp}$ in the five equations, E_{exp} being obtained from experimental B_n^m.

It should be noted, as shown already by Jørgensen, that this model holds good for these compounds. Thus one can think that σ-bonding may nearly be preponderant.

We see also that the error ΔE increases from Gd to Lu, for all the compounds: when Eu^{3+} takes the place of the rare-earth, the perturbation is the largest for Lu^{3+} and this deformation is not taken into account. It is the same for Nd^{3+} in YVO_4.

Using these values found for σ_0^* and the power law we can deduce, as it will further be seen, radial parameter σ_1 and σ_2 for the two distances R_1 and R_2 of the oxygen ligands.

V.2. Direct Determination of σ_1 and σ_2.

We want to test our intrinsic parameters calculated in section IV. So we have used them to determine directly the radial parameters σ_1 and σ_2 for each ligand series.

Let us take the previous example of the B_2 representation; the contributions may be separated and written:

$$4\frac{\sqrt{7}}{2}(5\cos^3\theta_1 - 3\cos\theta_1)^2\,\sigma_1 = \frac{8}{15}K_2^0(1)\,\overline{A}_2(R_1) + \frac{16}{11}K_4^0(1)\,\overline{A}_4(R_1) + \frac{1600}{429}K_6^0(1)\,\overline{A}_6(R_1)$$

$$4\frac{\sqrt{7}}{2}(5\cos^3\theta_2 - 3\cos\theta_2)^2\,\sigma_2 = \frac{8}{15}K_2^0(2)\,\overline{A}_2(R_2) + \frac{16}{11}K_4^0(2)\,\overline{A}_4(R_2) + \frac{1600}{429}K_6^0(2)\cdot\overline{A}_6(R_2)$$

as $B_n^m = K_n^m(1)\,\overline{A}_n(R_1) + K_n^m(2)\,\overline{A}_n(R_2)$

V.2.1. $\overline{A}_2$ Determination.

In section IV, the $\overline{A}_2$ values were not calculated. In fact, we have only one equation for two unknown quantities. It is thus necessary to assume a further relation (acceptable from a physical point of view): if we seppose that the σ effects are preponderant (other effects, π, δ, φ, vanish), the intrinsic parameters ratios have special values which can be calculated from *Kibler's* papers (*23, 24*).

$$\overline{A}_2/\overline{A}_4 = 4 \times 5/9$$
$$\overline{A}_2/\overline{A}_6 = 8 \times 5/13$$

If the two values of $\overline{A}_2$ calculated by these equations are close to each other, our hypothesis (σ effect dominant) is once more confirmed.

Doing this calculation, we have noted that the σ effects are dominant only for one series of ligands, either series 1 or 2 (corresponding to R_1 and R_2 distances).

Results are brought together in Table 15. It can be seen that for all the compounds of group I ($\overline{A}_4$ found by Newman's method), the ligands of series 2 have a σ-preponderant effect while for group II (power law canceled) the σ effect is dominant for the ligands of series 1. For the peculiar case of the arsenates (Eu^{3+}) intermediary between phosphates and vanadates, the two series of ligands give almost dominant σ effects.

For the two groups, in order to find the $\overline{A}_2$ of the nonpreponderant series, we start from the $\overline{A}_2$ already known and we use a power law since the $\overline{A}_2$ parameters have a more important electrostatic contribution than the $\overline{A}_4$ or $\overline{A}_6$. Effectively, if we consider the arsenates for which $\overline{A}_2(1)$ and $\overline{A}_2(2)$ may both be determined by the above mentioned formula, it may be noted that the calculation of t_n gives: $\overline{A}_2(1)/\overline{A}_2(2) = (R_2/R_1)^{5,5}$; 5,5 is close to 3, the value which we would have if the field was due to point charges.

In this way, for the europium ion in the vanadates more covalent than the arsenates, we have chosen a power $\geqslant 5,5$ about 10 and for the phosphates (more ionic) a power equal to 4. As for the vanadate differently doped, we have taken a value of 10 for Nd^{3+} which is close to Eu^{3+}, and for Er^{3+} and Tm^{3+} (which belong to *group* I) a value of 7.

Besides, if we calculate B_2^0 by the equation $B_2^0 = K_2^0(1)\,\overline{A}_2(1) + K_2^0(2)\,\overline{A}_2(2)$, we find again the phenomenological B_2^0 only for the vanadates doped with Eu^{3+} and for $YVO_4{:}Nd^{3+}$. All these compounds belong to *group* II (the most covalent compounds).

V.2.2. Results

With all these $\overline{A}_2$, $\overline{A}_4$, and $\overline{A}_6$ intrinsic parameters, the separated σ_1 and σ_2 contributions may be calculated. For each series of ligand we can determine a value of σ which satisfies all five equations as shown above.

The results are listed in Table 16, the values σ_1^* and σ_2^* found from σ_0^* (Jørgensen model) are also noted. It can be seen, chiefly for the ligands with σ effect dominant, that the agreement is quite good for these values.

It seems, then, that the choice of our method is justfied, since the instrinsic parameters are adequate, and more especially as the method used for compounds of *group* I is inadequate for compounds of *group* II and vice versa.

Table 15. $\bar{A}_2$ determinations (cm^{-1}).

	GdAsO$_4$	GdVO$_4$
Eu^{3+}	$\bar{A}_2(R_1)\ \left\{\begin{matrix}73\\105\end{matrix}\right. \rightarrow 89 \pm 14$	$\bar{A}_2(R_1)\ \left\{\begin{matrix}112\\88\end{matrix}\right. \rightarrow 100 \pm 12$
	$\bar{A}_2(R_2)\ \left\{\begin{matrix}52\\88.5\end{matrix}\right. \rightarrow 70.5 \pm 18$	$\bar{A}_2(R_2) = 62$
	$t_2 = 5.5$	$t_2 \sim 10$
YPO$_4$	**YAsO$_4$**	**YVO$_4$**
$\bar{A}_2(R_1) = 100$	$\bar{A}_2(R_1)\ \left\{\begin{matrix}89\\95\end{matrix}\right. \rightarrow 92 \pm 3$	$\bar{A}_2(R_1)\ \left\{\begin{matrix}98\\104\end{matrix}\right. \rightarrow 101 \pm 3$
$\bar{A}_2(R_2)\ \left\{\begin{matrix}86\\94\end{matrix}\right. \rightarrow 90 \pm 4$	$\bar{A}_2(R_2)\ \left\{\begin{matrix}60.5\\84.5\end{matrix}\right. \rightarrow 72.5 \pm 12$	$\bar{A}_2(R_2) = 55$
$t_2 = 4$	$t_2 = 5.5$	$t_2 = 10$
LuPO$_4$	**LuAsO$_4$**	**LuVO$_4$**
$\bar{A}_2(R_1) = 100$	$\bar{A}_2(R_1)\ \left\{\begin{matrix}90\\106\end{matrix}\right. \rightarrow 98 \pm 8$	$\bar{A}_2(R_1)\ \left\{\begin{matrix}104\\102\end{matrix}\right. \rightarrow 103 \pm 1$
$\bar{A}_2(R_2)\ \left\{\begin{matrix}98.5\\78.8\end{matrix}\right. \rightarrow 88.5 \pm 10$	$\bar{A}_2(R_2)\ \left\{\begin{matrix}56.5\\87.5\end{matrix}\right. \rightarrow 72 \pm 16$	$\bar{A}_2(R_2) = 50$
$t_2 = 4$	$t_2 = 5.5$	$t_2 = 10$

Tm^{3+}	Er^{3+}	Eu^{3+}	Nd^{3+}
$\bar{A}_2(R_1) = 104$	$\bar{A}_2(R_1) = 118$	$\bar{A}_2(R_1)\ \left\{\begin{matrix}98\\104\end{matrix}\right. \rightarrow 101 \pm 3$	$\bar{A}_2(R_1)\ \left\{\begin{matrix}157\\195\end{matrix}\right. \rightarrow 176 \pm 19$
$\bar{A}_2(R_2)\ \left\{\begin{matrix}79\\57\end{matrix}\right. \rightarrow 68 \pm 11$	$\bar{A}_2(R_2)\ \left\{\begin{matrix}84\\66\end{matrix}\right. \rightarrow 75 \pm 9$	$\bar{A}_2(R_2) = 55$	$\bar{A}_2(R_2) = 97$
$t_2 = 7$	$t_2 = 7$	$t_2 = 10$	$t_2 = 10$

YVO$_4$

Table 16. σ values (cm^{-1}).

			$GdAsO_4$		$GdVO_4$			
			(1)	(2)	(1)	(2)		
			σ^* 37	26	43	28		
			σ 32	32	42	29		
	YPO_4		$YAsO_4$		YVO_4			
	(1)	(2)	(1)	(2)	(1)	(2)		
Eu^{3+}	σ^* 42	34	41	28	42	26		
	σ 42	37	36	32	41	30		
	$LuPO_4$		$LuAsO_4$		$LuVO_4$			
	(1)	(2)	(1)	(2)	(1)	(2)		
	σ^* 44	33	43	27	44	25		
	σ 49	33	38	33	42	28		

	Tm^{3+}		Er^{3+}		Eu^{3+}		Nd^{3+}	
	(1)	(2)	(1)	(2)	(1)	(2)	(1)	(2)
YVO_4	σ^* 45	28	48.5	30	42	26	63	39
	σ 46	25	51	28	41	30	66	40

It seems interesting to use these parameters in a more extensive calculation involving effects other than σ ones. When σ effects are dominant this method not only accounts for them, but also gives a more reliable σ value since it includes the other effects quantitatively; this allows us to establish a comparison between the different compounds.

V.3. Kibler's Model (25).

Kibler, by a comparison of the point charge electrostatic model and the angular overlap model, has shown the equivalence between these two models; he has given some simple relations between intrinsic parameters for a series of ligands and antibonding radial contributions e_λ ($\lambda = \sigma, \pi, \delta, \varphi$) (Table 17). Kibler's I_n are different from our $\bar{A}_n$ (the basis is not the same). In Table 17 transformation formulas will be noticed. Kibler's e_σ is besides seven times greater than the σ^* found by Jørgensen's method.

The I_0 parameter determination method is analogous to the one used for the $\bar{A}_2$ parameter, using the hypothesis of a dominant e_σ effect. So we have:

$$I_0 = I_2/5 = I_4/9 = I_6/13 .$$

Table 17. Kibler's equations.

$$I_0 = \frac{1}{7}\,(\;\;e_\sigma + 2\,e_\pi + 2\,e_\delta + 2\,e_\varphi)$$

$$I_2 = \frac{5}{14}\,(\;2\,e_\sigma + 3\,e_\pi \qquad - 5\,e_\varphi)$$

$$I_4 = \frac{3}{7}\,(\;3\,e_\sigma + \;\;e_\pi - 7\,e_\delta + 3\,e_\varphi)$$

$$I_6 = \frac{13}{70}\,(10\,e_\sigma - 15\,e_\pi + 6\,e_\delta - \;\;e_\varphi)$$

$$I_2 = 2\,\bar{A}_2$$
$$I_4 = 8\,\bar{A}_4$$
$$I_6 = 16\,\bar{A}_6$$

$$e_\sigma = I_0 + \frac{4}{15}I_2 + \frac{6}{33}I_4 + \frac{100}{429}I_6$$

$$e_\pi = I_0 + \frac{3}{15}I_2 + \frac{1}{33}I_4 - \frac{75}{429}I_6$$

$$e_\delta = I_0 + \qquad - \frac{7}{33}I_4 + \frac{30}{429}I_6$$

$$e_\varphi = I_0 - \frac{5}{15}I_2 + \frac{3}{33}I_4 - \frac{5}{429}I_6$$

For each series of ligands, the average of the three I_0 values can be obtained. Let us note that the I_0 belonging to the ligands with e_σ predominant is calculated with a fair degree of precision, a fact which is another justification of our hypothesis. The I_0 for the other series of ligand is not so reliable.

Results:

The data are reported in Tables 18 and 19. In these tables are brought together the different e_λ values, calculated from Table 17. We can also find R values for each series of ligands, Jørgensen's σ^*, the σ obtained separately and the $E_\sigma = e_\sigma/7$ values.

Remarks common to both these tables can be made:

– We ascertain well that the e_σ contribution is always greater than the three other contributions, whatever the series considered. For the series where an e_σ effect is predominant, the other contributions are found to be quite negligible. For the other series of ligands, only the vanadates (Eu^{3+}), and mainly $YVO_4(Nd^{3+})$, give e_π, e_δ and e_φ contributions becoming important. This seems consistent since the second oxygen series in the vanadates are the farthest from the rare-earth, so they may more easily have bonds in other directions.

– For each compound, it can also be verified that the three values σ^*, σ, and E_σ are about the same for the series of ligands where the e_σ is dominant. (E_σ is the more reliable value.)

Table 18. e_λ values (cm^{-1}).

$LnMO_4$		PO_4		AsO_4		VO_4	
Gd	R_A			2.346	2.446	2.336	2.462
	e_σ			258	207	273	155
	e_π			– 16	– 19	12	– 29
	e_δ			18	20	– 13	42
	e_φ			– 6	– 7	5	– 25
	σ^*			37	26	43	28
	σ			32	32	42	29
	E_σ			37	30	39	22
Y	R_A	2.313	2.374	2.300	2.412	2.291	2.433
	e_σ	244	254	259	210	283	128
	e_π	26	– 4	– 2	– 12	4	– 45
	e_δ	– 19	4	3	13	– 4	71
	e_φ	1	– 2	– 1	– 1	2	– 37
	σ^*	42	34	41	28	42	26
	σ	42	37	36	32	41	30
	E_σ	35	36	37	30	40	18
Lu	R_A	2.264	2.346	2.248	2.385	2.242	2.408
	e_σ	246	240	279	210	288	133
	e_π	39	8	– 8	– 16	2	– 41
	e_δ	– 41	– 11	8	18	– 1	57
	e_φ	12	6	– 3	– 6	1	– 27
	σ^*	44	33	43	27	44	25
	σ	47	33	38	33	42	28
	E_σ	35	34	40	30	41	19

Table 19. e_λ values for YVO_4 (cm^{-1}).

	Tm^{3+}		Er^{3+}		Eu^{3+}		Nd^{3+}	
	(1)	(2)	(1)	(2)	(1)	(2)	(1)	(2)
e_σ	281	184	287	212	283	128	503	134
e_π	22	11	40	9	4	– 45	– 20	– 51
e_δ	– 25	– 12	– 36	– 10	– 4	71	21	122
e_φ	9	4	6	4	2	– 37	– 8	– 86
σ^*	45	28	49	30	42	26	63	39
σ	46	25	51	28	41	30	66	40
E_σ	40	26	41	30	40	18	72	19

For the other series, it seems consistent not to expect such a good concordance: the E_σ value is all the more different from σ^* and σ as the σ effect is less dominant.

Let us look again at Table 18. If we consider the compounds in which Eu^{3+} replaces the same rare-earth ion (horizontal line in the table), it will be noticed for the first series of ligands that the e_σ value increases from PO_4 to AsO_4 and to VO_4; on the other hand, e_σ decreases from PO_4 to VO_4 for the second series. But we see that the R_1 values decreases from PO_4 to VO_4 while the R_2 values increase. The e_σ parameter always increases and this is true for the three types of compounds: gadolinium, yttrium, and lutetium.

This is obviously correct since e_σ is a radial parameter related to the orbital overlap, becoming larger as the distances become smaller.

Now let us consider Table 19 again when Y^{3+} is replaced by Tm^{3+}, Er^{3+}, Eu^{3+}, and Nd^{3+}; the e_σ values of the first series of ligands do not vary significantly from Tm^{3+} to Eu^{3+}, but for Nd^{3+} the e_σ contribution is clearly more important; it is probably due to a larger orbital overlap connected with the greater 4 f shell radius of

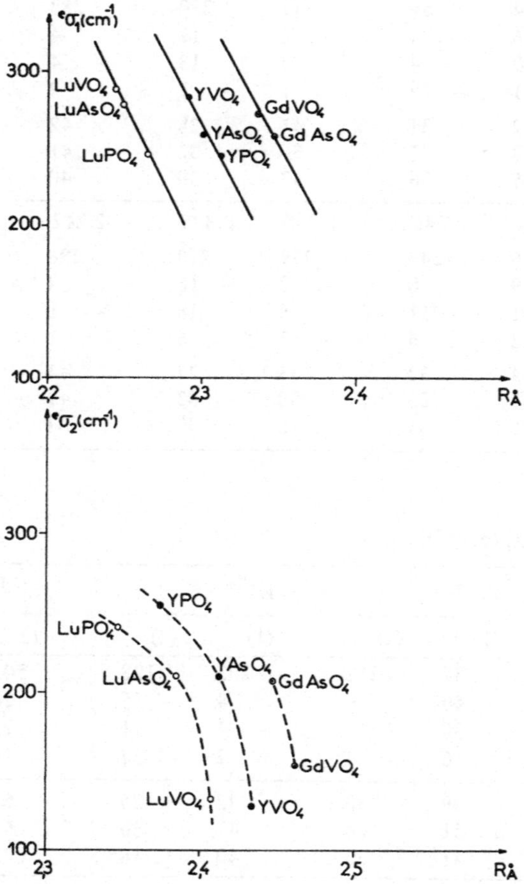

Fig. 2.

this ion. This result is interesting because it shows that e_σ is more sensitive to the size of the rare-earth included in the bonding than to the ligand – rare-earth distance which differ little from Tm^{3+} to Nd^{3+}.

To show how e_σ is connected with the distance, Figure 2 presents the e_σ values plotted as a function of R for each series of ligands.

For the ligands of series 1 (Fig. 2a) three curves which seems parallel straight lines. On each single line we find the compounds of one rare-earth ion, in the order VO_4, AsO_4, PO_4. It may be assumed that the slope is connected with the ionicity variation from VO_4 to PO_4. Moreover when Eu^{3+} replaces the rare-earth, the perturbation is different from Lu to Y and Gd, which explains the separation of the straight lines.

We may also see that, on the hand, the e_σ decrease observed on any one line from VO_4 to AsO_4 and PO_4, is important in the three cases; on the other hand, with the vanadates, for example, the e_σ decrease observed from Lu to Gd is less noticeable. The same remark holds for the arsenates or phosphates. Effectively R increases from PO_4 to VO_4 for one rare-earth but less than inside a single MO_4 group, from Lu to Gd. The parameter e_σ is thus more dependent on the covalency than on the distance.

For the ligands of series 2 (Fig. 2b), the results are less clear; nevertheless a rough classification can be established.

VI. Concluding Remarks

This work has allowed us to see the evolution of the rare-earth – ligand bonding nature inside the $LnMO_4$ series.

The comparison between the phenomenological parameters calculated from optical spectra and the electrostatic point charge parameters has shown that the crystal field is more different from an electrostatic one than is usual; but a ligand field hypothesis seems reasonable.

It has been observed that Newman's method for determining intrinsic parameters $\bar{A}_n$ can be used for classifying the compounds according to their covalency degree.

We have thus established two groups:

– a first group including the less covalent compounds $LnPO_4(Eu^{3+})$, $LnAsO_4(Eu^{3+})$, and $YVO_4(Tm^{3+}$ or $Er^{3+})$: in this case Newman's method is applied only for the $\bar{A}_4$ calculation, as usual with a power law; the crystal field thus keeps its electrostatic features.

– a second group including the more covalent compounds $LnVO_4(Eu^{3+})$ and $YVO_4(Nd^{3+})$: a generalized method canceling the power law was used here; thus the electrostatic feature of the field vanishes.

We have also found that in the one compound YVO_4, the different activators did not belong to the same group. The perturbation introduced by a rare-earth larger than Y^{3+} (Eu^{3+} or Nd^{3+}) modifies the crystal field, removing its electrostatic features. This different behavior can be found again and explained by A_n^m considerations.

From phenomenological parameters an angular overlap calculation according to the Jørgensen's method points out the σ-bonding dominance in all the studied compounds.

Moreover a σ calculation for each ligand series, directly using intrinsic parameters has justified the choice of these parameters.

We have then been able to apply Kibler's model involving all the effects: σ, π, δ, φ; this model requires the knowledge of intrinsic parameters.

We have confirmed that the e_σ effects were generally dominant; in the case of europium-activated vanadates and $YVO_4(Nd^{3+})$, the e_π, e_δ, and e_φ contributions for the second series of oxygens may be important: these oxygens are the farthest from the rare-earth and can be bonded in other directions.

We have also noticed an increase of the e_σ values with a decrease of the R distance rare-earth-oxygen. Moreover, the e_σ values seem more sensitive to the nature of the group MO_4 (P, As, or V) than to the nature of the host cation (Gd, Y, or Lu). Besides, in the one compound YVO_4, diversely doped and in which the distance rare-earth-oxygen differs little from one ion to another, the greatest e_σ contribution was found for Nd^{3+}. From these results it can be concluded that e_σ is more dependent on covalency than on the distance.

Thus, from a large number of compounds doped with rare-earth, optical studies can bring interesting information on the rare-earth ligand bonding.

Acknowledgments: We are grateful to Professor C.K. Jørgensen for interesting discussion concerning this paper.

References

1. *Strada, M., Schwendimann, G.:* Gazz. Chim. Ital. **64**, p. 662 (1934).
2. *Koster, G.F., Dimmock, J.O., Wheeler, R.G., Statz, H.:* Properties of the thirty-two point groups. Cambridge: MIT Press, 1963.
3. *Schwarz, von H.:* Z. Anorg. Allg. Chem. **323**, p. 44 (1963).
4. *Lohmüller, von G., Schmidt, G., Deppisch, B., Gramlich, V., Scheringer, C.:* Acta Cryst. B **29**, p. 141 (1973).
5. *Baglio, J.A., Sovers, O.J.:* J. of Solid State Chem. *3*, p. 458 (1971).
6. *Newman, D.J., Urban, W.:* J. Phys. C: Solid State Phys. **5**, p. 3101 (1972).
7. *Brecher, C., Samelson, H., Riley, R., Lempicki, A.:* J. Chem. Phys. **49**, p. 3303 (1968).
8. *Brecher, C., Samelson, H., Lempicki, A., Riley, R., Peters, T.:* Phys. Rev. **155**, p. 178 (1967).
9. *Durif, A., Forrat, F.:* C.R. Acad. Sc. Paris, **245**, p. 1636 (1957).
10. *Aia, M.A.:* J. Electrochem. Soc. **114**, p. 367 (1967).
11. *Linares, C., Louat, A.:* J. Phys. **36**, p. 717 (1975).
12. Quantum Chemistry Program Exchange n° 216.
13. *Karayianis, N., Morrison, C.A., Wortman, D.E.:* J. Chem. Phys. **62**, p. 4125 (1975).
14. *Kuse, D.:* Z. für Phys. **203**, p. 49 (1967).
15. *Wortman, D.E., Leavitt, R.P., Morrison, C.A.:* J. Phys. Chem. Solids **35**, p. 591 (1974).
16. *Hutchings, M.T.:* Sol. State Phys. **16**, p. 237 (1964).
17. *Louat, A., Linares, C., Blanchard, M.:* C.R. Acad. Sc. Paris, **279**, série B, p. 63 (1974).
18. *Newman, D.J.:* Adv. Physics **20**, p. 197 (1971).
19. *Newman, D.J.:* Proceedings of the 10th rare earth research Conference Carefree U.S.A., p. 1135 (1973).
20. *Freeman, A.J., Watson, R.E.:* Phys. Rev. **127**, p. 2058 (1962).
21. *Schäffer, C.E., Jørgensen, C.K.:* Mol. Phys. **9**, p. 401 (1965).
22. *Kuse, D., Jørgensen, C.K.:* Chem. Phys. Lett. **1**, p. 314 (1967).
23. *Kibler, M.R.:* Chem. Phys. Lett. **8**, p. 142 (1971).
24. *Kibler, M.R.:* J. Chem. Phys. **55**, p. 1989 (1971).
25. *Kibler, M.R.:* Int. J. Quantum Chem. **9**, p. 403 (1975).

Author-Index Volume 1—33

Topics in Current Chemistry

Fortschritte der chemischen Forschung

Managing Editor:
F. L. Boschke

Volume 58
New Theoretical Aspects
60 figures, 7 tables. IV, 186 pages. 1975
ISBN 3-540-07355-8

Contents: On the Description of Molecules Using Point Charges and Electric Moments. – CRAMS – An Automatic Chemical Reaction Analysis and Modeling System. – IR Fourier Transform Spectroscopy.

Volume 60
Structure of Liquids
88 figures, 38 tables. IV, 205 pages. 1975
ISBN 3-540-07484-8

Contents: Molecular Models for the Solvation of Small Ions and Polar Molecules. – Conjectures on the Structure of Amorphous Solid and Liquid Water.

Volume 61
Physical and Chemical Applications of Dyestuffs
96 figures, 23 tables. IV, 187 pages. 1976
ISBN 3-540-07559-3

Contents: Organic Dyes in Laser Technology. – Reaction of Excited Dye Molecules at Electrodes. – Application of the Semiconductor Properties of Dyes Possibilities and Problems. – Organic Dyestuffs as Catalysts for Fuel Cells.

Volume 63
Bonding and Structure
55 figures, 25 tables. IV, 202 pages. 1976
ISBN 3-540-07605-0

Contents: Discriminating Interactions Between Chiral Molecules. – No-Bond-Resonance Compunds, Structure, Bonding and Prosperties. – The Molecular Zeeman Effect.

Volume 65
Theoretical Inorganic Chemistry II
47 figures, 44 tables. IV, 153 pages. 1976
ISBN 3-540-07637-9

Contents: Diastereoisomerism and Diastereoselectivity in Metal Complexes. – Mechanistic Aspects of the Photochemical Reactions of Coordination Compounds. – Complexation and Activation of Diazenes and Diazo Compounds by Transition Metals.

Springer-Verlag
Berlin
Heidelberg
New York

NMR
Basic Principles and Progress
Grundlagen und Fortschritte

Editors: P. Diehl, E. Fluck, R. Kosfeld

Volume 10

Van der Waals Forces and Shielding Effects

13 figures, 46 tables. II, 118 pages. 1975

ISBN 3-540-07340-X

Contents: Van der Waals Forces in NMR Intermolecular Shielding Effects: Historical Development (up to 1961). – Continuum Models. – Pair Interaction Models σ_w. – Other Experimental Proton Data on σ_w. – The Physical Nature of the Field F^2 and of the Associated Excitation Energy. – The Site Factor. – The Repulsion Effect. – The Effects of Higher Order Dispersion Terms. – The Parameters B. – σ_w in Dense Media. – The Temperature Dependence of σ_w. – Factor Analysis. – $^{19}F\sigma_w$ Studies. – σ_w of Nuclei other than ^1H and ^{19}F. – Alternate Referencing Systems. – On the Required Molecular Parameters and Physical Constants.

Volume 11: M. Mehring

High Resolution NMR Spectroscopy in Solids

104 figures. XI, 246 pages. 1976

ISBN 3-540-07704-9

Contents: Introduction. – Nuclear Spin Interactions in Solids. – Multiple-Pulse NMR Experiments. – Double Resonance Experiments. – Magnetic Shielding Tensor. – Spin-Lattice Relaxation in Line Narrowing Experiments. – Appendix.

Volume 12: B. Lindman, S. Forsén

Chlorine, Bromine and Iodine NMR

Physico-Chemical and Biological Applications

74 figures, 29 tables. XIV, 368 pages. 1976

ISBN 3-540-07725-1

Contents: Introductory Aspects. – Relaxation in Molecules or Ions with Covalently Bonded Halogens. – Shielding Effects in Covalent Halogen Compounds. – Scalar Spin Couplings. – Relaxation of Chloride, Bromide and Iodide Ions. – Shielding of Halide Ions. – Quadrupole Splittings in Liquid Crystals. – Halide Ions in Biological Systems. – Studies of the Perchlorate Ion.

Volume 13:

Introductory Essays

Editor: M. M. Pintar

48 figures. XI, 154 pages. 1976

ISBN 3-540-07754-5

Contents: A Guide to Relaxation Theory. – Thermodynamics of Spin Systems in Solids. – Coherent Averaging and Double Resonance in Solids. – Macroscopic Dipole Coherence Phenomena. – Nuclear Spins and Non Resonant Electromagnetic Phenomena. – Nuclear Spin Relaxation in Molecular Hydrogen. – Longitudinal Nuclear Spin Relaxation Time Measurements in Molecular Gases. – Spin-Lattice Relaxation in Nematic Liquid Crystals Via the Modulation of the Intramolecular Dipolar Interactions by Order Fluctuations. – NMR Studies of Molecular Tunnelling. – Effect of Molecular Tunnelling on NMR Absorption and Relaxation in Solids. – How to Build a Fourier Transform NMR Spectrometer for Biochemical Applications.

Springer-Verlag
Berlin
Heidelberg
New York

GPSR Compliance

The European Union's (EU) General Product Safety Regulation (GPSR) is a set of rules that requires consumer products to be safe and our obligations to ensure this.

If you have any concerns about our products, you can contact us on ProductSafety@springernature.com

In case Publisher is established outside the EU, the EU authorized representative is:

Springer Nature Customer Service Center GmbH
Europaplatz 3
69115 Heidelberg, Germany

Batch number: 10613743

Printed by Printforce, the Netherlands